BUILDING ON BORROWED TIME

BUILDING ON
Rising Seas and Failing
BORROWED
Infrastructure in Semarang
TIME

LUKAS LEY

University of Minnesota Press | Minneapolis | London

Excerpt from Theodoros Chiotis, "Perfusion," in *Futures: Poetry of the Greek Crisis,* ed. and trans. Theodoros Chiotis (London: Penned in the Margins, 2015), reprinted by permission of Theodoros Chiotis.

Portions of chapter 3 appeared in "On the Margins of the Hydrosocial: Quasi-Events along a Stagnant River," *Geoforum* (April 2018), used with permission from Elsevier. Portions of chapter 4 appeared as "Figuring (Out) the Sinking City: Tidal Floods and Urban Subsidence in Semarang, Indonesia," in *Disastrous Times: Beyond Environmental Crisis in Urbanizing Asia,* ed. Eli Elinoff and Tyson Vaughan (Philadelphia: University of Pennsylvania Press, 2020); copyright 2020 University of Pennsylvania Press.

Copyright 2021 by the Regents of the University of Minnesota

All rights reserved. No part of this publication may be reproduced, stored in a retrieval system, or transmitted, in any form or by any means, electronic, mechanical, photocopying, recording, or otherwise, without the prior written permission of the publisher.

Published by the University of Minnesota Press
111 Third Avenue South, Suite 290
Minneapolis, MN 55401-2520
http://www.upress.umn.edu

ISBN 978-1-5179-0887-4 (hc)
ISBN 978-1-5179-0888-1 (pb)
Library of Congress record available at https://lccn.loc.gov/2021025919.

The University of Minnesota is an equal-opportunity educator and employer.

CONTENTS

Introduction: Tidal Flooding and Chronic Infrastructural Breakdown   1

1. Becoming: Semarang's Swamp in Late Colonial Times   33

2. Stuck: Never-Ending River Normalization   57

3. Floating: Endurance and the "Quasi-Events" of Living with Flooding   85

4. Figuring: Environmental Governance and the Political Affordances of Infrastructure   121

5. Promise: Remodeling Drainage   151

Afterword   191

ACKNOWLEDGMENTS   197

GLOSSARY   201

NOTES   205

REFERENCES   215

INDEX   227

INTRODUCTION
TIDAL FLOODING AND CHRONIC INFRASTRUCTURAL BREAKDOWN

> The perfect grip of a hand
> that is no longer of any use
> landscapes the present
>
> —THEODOROS CHIOTIS, "Perfusion"

THE RAINS LET UP shortly before the event was scheduled to begin. The moisture was sure to have a pleasant cooling effect on the day, creating perfect conditions for Kemijen's healthy living event, *jalan santai* (relaxed walking). When I arrived, the left bank of the Banger River was packed with residents dressed in athletic wear, readying to begin. I was nudged along to the starting line by friends. The moment I reached it, the signal sounded, and we were set in motion. Two men posted behind the starting line distributed lottery tickets and snack coupons as an additional incentive to participate. Walking side by side, Ariel and I enjoyed the opportunity for a chat.[1] She was in a good mood and quite talkative compared to her usual taciturn self. Having grown up in Kemijen, Ariel knew its geography and neighborhoods well. Many of its residents frequented her makeshift canteen by the river, which afforded her a deep knowledge of local developments and regular gossip. As we progressed, she offered commentary on shifts in our surroundings. The riverside landscape was suddenly dipped in deep, earthy, brownish tones; the unpaved streets were covered in dry mud; and, in contrast to more affluent areas, the number of wooden structures far outweighed those made of concrete. "This is where the dirty part begins," she informed me. It would have been hard to miss the change. The ruins of many houses, rotting in a stagnant brew of algae and shit, populated the area like sad, half-forgotten sculptures. While the whole neighborhood experienced regular flooding, this area was particularly affected because it bordered an open sewage canal prone to overflowing. As we continued, I asked Ariel when her family would raise the front part of their house—I knew that their living room flooded regularly during high tide despite being located in a "cleaner" area. She replied that they had to be patient and that it would be smarter to first remove the front part of their home where her twin sister ran a small cell-phone credit business. That part, so she reasoned, narrowed the wastewater conduit, preventing it from releasing used water back into the river.

Wastewater and runoff are supposed to collect in the Banger River before being channeled to and pumped into the nearby ocean. But this century-old drainage infrastructure has long stopped working. At present, it is kept in a barely functional state by multiple government and community initiatives. Up and down the river, neighborhood groups and residents employ diesel pumps or other improvised techniques to prevent neighborhood-level flooding and continuously reinstate flow despite repeated technical fixes of the river infrastructure. And yet these combined efforts are insufficient, which is why Kemijen and other parts of North Semarang still regularly flood with putrid, contaminated river water. Toxic water and exhaustion permeate the neighborhood of Kemijen and extensive parts of Semarang's floodplain. Seawater mixes with runoff and wastewater, resulting in all kinds of infrastructural transgressions and excess. Water appears outside engineered conduits (rivers, gutters, and channels), pressing up from underneath streets and tiles or seeping through riverbanks. The labors of pumping, warding off, cleaning, and fighting waterborne diseases all inevitably lead to bodily exhaustion. Nevertheless, the Sisyphean fight for regular flow—of water and sewage—is an important and necessary activity in Semarang. Ariel's husband, Arief, who in 2012 initiated a community-run water pump group (*pompanisasi*) at the peak of tidal flooding to prevent the gutter between their houses and the street from flooding their home, daily dedicates time and energy to maintain a hydraulic self-help system that channels wastewater into the Banger. His family, like so many others, is worried about the future, since their house is located on land that is steadily sinking. In addition to enabling flow, they also regularly have to lift the foundations of their houses, fighting a tiring battle against gravity and breakdown. Submersion is imminent and recurrent in the environment that they and thousands of other families inhabit.

Many neighborhoods in Semarang's North are built on former wetland and rest on human-made soils currently sinking at a worrisome rate of ten to fifteen centimeters per year. The landscape is stitched through by the capillaries of a water infrastructure now rendered largely dysfunctional. From this complex nexus of liquids, materials, and flood-prevention practices originates a common but unequally distributed threat: toxic river water overflowing from drains and inundating peoples' houses, liable to cause a variety of waterborne diseases and damage personal belongings. At the worst of times, this brackish wastewater remains trapped in homes and gutters instead of flowing into the central drain—the Banger River. This looming danger defines the present and near future, as residents are forced to continuously devise strategies, improvise short-term solutions, and plan in light of imminent flooding. Events of flooding, however, rarely disrupt daily rhythms; they simply occasion fixes and regularized pumping, wiping, and sweeping. In this book, I argue that all these preventive efforts are nested within a specific temporality: the chronic present.

Floodplain residents who live in this chronic present must deal with the constant possibility of infrastructural breakdown and actual events of technological failure. This experience of time shapes people's relationships with the future in specific ways. It socializes floodplain populations into lives governed by the futurity of breakdown and recurrent flooding. Most residents respond to this present by repairing and rebuilding domestic and public infrastructures. While flooding itself as well as other water-related problems are, then, essential features of this chronic present, this relation to time is not just an effect of the inherently moist environment that people live in. That is, it is also a product of a series of political exclusions that can be traced back to colonial times, and which created a spatialized moral hierarchy of populations, placing the residents of Semarang's coastal area at its bottom. Today, flood victims are stuck in a world concerned with presentness because long-term fixes always flatline, changing little in the way of providing people with actionable future plans. Understanding this time is particularly important at a moment when Dutch water experts are returning to Indonesia and Southeast Asia to pursue water management projects. This Dutch polder (hydraulic system) building mania follows a technofix mentality that perpetuates the longue durée experience of water governance instead of setting the direction for wholesale social change. In fact, the Dutch flood mitigation pilot project presented in this book temporarily fixed flooding at the cost of reproducing a politically stagnated situation in which poor floodplain residents are still treading water.

Rob

When Semarang's floodplain rivers and tributaries swell with oceanic tidewater, the drains of coastal neighborhoods are prone to clog, like choked veins, resulting in larger infrastructural failure, such as seeping riverbanks and oozing house floors. In Semarang, people call this phenomenon *rob*. Because levels of inundation differ from household to household and from neighborhood to neighborhood, events of *rob* have diverse consequences. Though many people in several contiguous subdistricts might agree that the tide (*air pasang*) is particularly high on a given day, that does not mean that people will be affected by *rob* to the same degree. Therefore, there is no overarching logic of flooding. Diverse temporal, spatial, and social arrangements are tantamount, such as when the neighborhood's streets were last raised (*diuruk*), when a local employer last provided funds to patch up a riverbank, or how much water a government-built retention basin can absorb. As Ariel noticed during our walk through Kemijen, levels of flood exposure break down at the scale of the neighborhood. For example, a wealthy city council member who was the main sponsor of the *jalan santai* event related at the beginning of this chapter recently built a luxurious, elevated villa amid molding houses on the west bank, while Ariel and her husband scramble to find funds to renovate their flood-stricken home (at the added expense of potentially dismantling her sister's business space). Their house is old, and its ground floor has sunk well below street level. And there are yet others whose houses flood *every* night; these are the houses of those living in pockets of Kemijen that Ariel qualified as "dirty." Using small electric pumps, they pump the water out each night, knowing it will return with unforgiving predictability. Others, less fortunate, stack bricks or whatever material they can find against seeping riverbanks in an attempt to block overflow. Relief, damage, and exhaustion from unruly water flows are thus distributed inequitably, and people must resort to distinctive architectures of time (Sharma 2014) in order to respond to perceived risks. These orientations to the temporality of *rob* serve to respond to and act on the near future. They are tailored to this more immediate future because long-term governmental development plans for the future provide unactionable information since they rarely, if ever, materialize. In fact, the time horizons of such plans even disavow the continued existence of those who live with chronic flooding, envisioning large-scale relocations due to rising sea levels. Erik Harms (2013) has shown that eviction plans in Saigon form a temporal structure that undermines individual planning as a productive relationship with the future. Many residents experience urban redevelopment plans as a recurrent threat to their position in time and space, forcefully suspending them in the now. As a mechanism of control, temporal instability makes planning for the future virtually impossible for many of them, which is why they cling to the present. *Rob* has become a structuring principle in the lives of residents of Semarang's North as well, with the difference that it is not just the government's plans that upset their productive relationship with time but also the volatile environment of the swamp.

As I show in this book, flood management plans for the near future make up a complex set of interrelated temporalities. Individual and more collective strategies, coping mechanisms, and state-administered drainage technologies all comprise an endless "meantime" (Cazdyn 2012), or stretched-out present. This chronic present is characterized by recurrent risk of sinking and the constant need to responsibly self-manage flooding. These small, mostly uncoordinated acts of repair maintain the illusion of a viable future, one in which residents will be safe from *rob* and toxicity. It is a potential future that humans clasp to in many places all over the world, a future condemned to an ephemeral existence between the materialist promises of sovereigns and infrastructural projects that never were (Howe et al. 2016).

Rob, which has Ariel's family and scores of residents in North Semarang deeply concerned about their future, differs from the apocalyptic catastrophes that haunt the lives of many Indonesians: the gray clouds that threaten Jakarta during rainy seasons, the seismic power of Java's volcanoes that can unleash devastating tsunamis, or the mudslides that can kill dozens or even hundreds in a single swipe. Instead, *rob* is a force generated by the chronic failure of dated drainage infrastructure built and rebuilt on sinking land.

Building on Borrowed Time

When one broadens the view from Kemijen to the scale of the city, it is clear that the complex chronography of flood prevention and *rob* follows a spatial hierarchy nested in a (post)colonial imaginary of order. The Northeast has not benefited from the same infrastructural investments—such as drainage and housing improvement schemes—as centrally located and uptown neighborhoods. For example, Semarang's international airport is also located on marshland, but its private operator has recently implemented a sophisticated polder that allows the airport to process travelers year-round, uninterrupted by tidal or seasonal changes. The airport prides itself with operating the first floating terminal: its newest "eco-friendly" building is built on top of a swamp and supposed to preserve "the surrounding areas, including the beach, the soil and the water" (Rohmah 2013). By contrast, large parts of Semarang's coastal wetlands form a neglected territory, one that was largely disallowed the infrastructural fixes and futurities of postcolonial development. For the most part, current government plans for North Semarang continue to perpetuate an ideology that indexes the North as exhibiting inappropriate and illegitimate ways of adapting to ecological conditions. While the phenomenon of tidal flooding has in fact renewed state efforts to save floodplain communities from sinking into decay, these programs materialize unevenly and often fail because they disregard alternate causes of flooding, such as ground water extraction leading to land subsidence.

Therefore, less fortunate areas require discrete acts of synchronizing with and adapting to the operations of water infrastructure or lack thereof. When infrastructures become obsolete through neglect and underinvestment or new infrastructures override older ones, residents of these areas are caught out of sync and must scramble to catch up. They solve the problem of being "out of time" through economic, affective, and social investments, investments I call "building on borrowed time." The ongoing maintenance and repair of Semarang's Dutch-designed drainage system, a sociotechnical system that is flexing and leaking its way through time (S. Jackson 2014), constitutes a form of borrowing time. This time is borrowed from sociotechnical arrangements, the sinking floodplain, and governmental efforts to extend the life of infrastructure. But like all modern artifacts, this system has capacity limits and an expiration date. The crucial question is where individuals will be positioned, in time and space, with regard to the systems' breakpoints when its flexing and bending reaches its limit. In coastal cities, everyday life and infrastructural systems are entangled in ways that turn small breakages into cascades of effects. For example, in Jakarta, where I lived for several months in 2011 and 2014, flood safety depends on the largely uncoordinated operation of a loose network of dams and pumps. As Abidin Kusno (2018, 28) has shown, residents of the northern neighborhood Pluit "can never quite develop a securely self-contained infrastructure, no matter how many pumps are installed." In the North of Jakarta, sinking is a compound disaster caused by groundwater extraction and sea level rise that disproportionately hits the inhabitants of slums. Living with chronic flooding in cities like Jakarta or Semarang (Padawangi and Douglass 2015) means to depend on and negotiate the (in)visibility of infrastructures, and forces one to constantly interrogate one's position toward them (see Colven 2020).

For residents of Kemijen, enduring infrastructural failure requires loaning time in the form of infrastructural sociotechnical synchronization. This loan time also steals time from debtors by regularly cutting into their resources. The language of debt is a useful conceptual tool to grasp how bridging the gap between intentioned infrastructural development and existing conditions is experienced by floodplains residents. When Ariel and I contemplated the "dirty part" of Kemijen, we were witnessing a kind of bankruptcy, the ultimate impossibility of borrowing time experienced by some residents—to borrow and to build on borrowed time. They were evicted by *rob*. Still, some residents endured in houses that were basically submerged: soaked mud floors, mattresses and belongings sheltered on improvised stilts, defunct toilets. Based on Karl Marx, Shahram Khosravi (2019) has argued that bordering practices steal time from refugees who are made to wait in war ruins or camps. Similar to waiting, an effect of subordination experienced by less powerful groups in society, borrowing time puts one in an inferior position toward dominant groups of society.

In Semarang, one important mode of borrowing time is to level up, or *meninggikan*. The root word *tinggi*, which means high or tall, can become an active noun in Indonesian: *peninggian*. It then refers to the increase or elevation of either a material thing, like a road, or an immaterial thing, like temperature. Peninggian has become a bureaucratic keyword as well as an everyday category in Semarang and other coastal cities, such as Jakarta. *Peninggian* translates to elevating or raising. The word bears an interesting resemblance to and creates a productive tension with another category in Indonesian history and nationalist discourse, namely *pembangunan*. The meaning of pembangunan vacillates between construction and progress. At some point in Indonesian public discourse, these meanings became interdependent and even one and the same thing. The concept of pembangunan, which came to stand for development, was particularly dear to Indonesia's longtime authoritarian president, Suharto. Under his rule, it symbolized the ascent of Indonesia as a major economic player with global ambitions. While pembangunan, or development, has the connotation of progressively building a better future, peninggian, or elevating, is a mere technical fix that extends the life spans of public infrastructure. This elevating is synchronized with cycles of breakdown that it itself cannot prevent. Peninggian, for many residents, is a debt relation to land that they inherited from past generations. They consider it an ecological bond. Peninggian is often compared to "renting from nature," as I show in this book. This rent, however, is not necessarily a financial relation; residents combine public subsidies, donor money, and their own funds to make house renovations, but they also mobilize energy, time, effort, and social relations (see Kleinman 2014). The phrase, then, points to the ongoingness of infrastructural incompleteness, which leads to forms of economic and physical as well as emotional taxation.

Construction in Kemijen neither accrues wealth nor translates into veritable building in a modernist sense. Elevating houses and streets does not lead to progress (pembangunan) but captures both residents and landscape in a never-ending tango with the present. Their investments inevitably reproduce this present's temporal fixtures instead of setting the stage for change. Flooding constitutes a temporal regime that forces residents in North Semarang to engage in something that Nicolas Lainez fittingly calls "treading water." Lainez (2019, 806) uses this phrase to describe how precarious Vietnamese sex workers anxiously "put effort into keeping themselves afloat but never furthering their status and lives or catching up with the currents of development and progress." For this surplus population excluded from formal education and stable jobs, staying afloat requires risking their health and lives in highly precarious and dangerous work settings. Importantly, Lainez shows that "desynchronized" sex workers enjoy relative autonomy from the exigencies of clock time labor while engaging in present-oriented practices, such as gambling, which "distorts and shrinks the past and the future towards a saturated and speculative present" (814). Similar to Lainez, I hold that the poor population of Semarang's North is literally "submerged" in a presentness, a state of being reproduced by the environmental fluctuations of the intertidal zone in which they live and entrenched by the systematic infrastructural neglect and desynchronization of the city's North. While practices of "treading water" thrive on anxiety and transience, Lainez points out that they also generate pleasure and social attachment as well as empowerment. As I will show, peninggian can also generate pride and social belonging. Often, such acts of repair suggest at least some grip on this present and generate spaces from which to project oneself into a safer future, no matter how temporary this foundation may be. Building on borrowed time is akin to what Steven Jackson (2014, 222) has described as "arts of repair by which . . . sociotechnical forms and infrastructures, large and small, get not only broken but restored, one not-so-metaphoric brick at a time." Continual acts of repair and community-run pumps along the Banger River allow residents to survive in the aftermath of developmental interventions and projects of governmentality. While I share Jackson's fascination for these acts of repair and retrofit, what happens when this aftermath turns into a chronic present? Can we still marvel at the creative abilities of people, their ethics of care, when these acts lead nowhere? As I argue, the annual procedure of peninggian and other repetitive acts of repair capture the difficulty of imagining an otherwise, a structural inability that is symptomatic of Eric Cazdyn's "chronic present."

The Chronic Present

Thinking about Semarang's infrastructures in ruin through the lens of temporality, I have found inspiration in what Eric Cazdyn refers to as the new chronic. In his book *The Already Dead,* Cazdyn (2012, 14) considers time as "a privileged path to understanding our present moment." He argues that a new temporal gap has opened between "ever-shifting economic and political realities and the institutions and ideologies we have available to us to cope with these new realities—institutions and ideologies suited to another moment of capitalism and unfit for the present" (103). He also reveals processes by which institutionalized actors try to measure and shape time as the capitalist "world-system reconfigures [itself] and institutions, corporate, national, and ideological, struggle to keep pace" (101). For example, by affording the means to prolong and sustain life, medicine has successfully filled the gap between the present and the future. Based on medicine's advances in prophylactic and palliative treatment, Cazdyn argues that contemporary society has entered something like a new chronic mode (5). This mode of time suggests that breakdown and disease are something normal, even banal, and that they can be managed, especially if diagnosed in a timely fashion. This is a mode of time, according to Cazdyn, that cannot account for terminality or acuteness. In this chronic mode, the risk of crisis is calculable and manageable (see also Beck 1992; Dean 1998; de L'Estoile 2014). But, as Cazdyn and many others have illustrated, risk never disappears in the worlds created by neoliberal capitalism, even when powerful institutions try to manage it. Certainly, this point was brought home by the devastating COVID-19 virus, which reminded us that the question is not whether crisis will occur but *when.* Cazdyn's theorization of time as punctured by crisis also touches on the fragile nature of political ecologies. As a key domain of capitalist value production, he argues, ecologies are produced in ways that inevitably result in crisis and disaster. Crises only serve to strengthen and reproduce the capitalist system, which is programmed to be deficient. "Natural" disasters are quite predictable from the viewpoint of the chronic. As Cazdyn points out, experts knew exactly what would happen if the New Orleans levees broke. Instead of tackling systemic problems and obvious vulnerability, we learn to live with and normalize potential nuisance. Cazdyn associates this attitude toward disaster with dominant ideologies of life and death: "once the malady is identified and understood, we can adjust ourselves to its presence" (2012, 68). What capitalism needs is a cultural configuration that expects, anticipates, and plans for failure and breakdown. Crisis—in national health systems, regional economies, and planetary ecosystems—today merely demands of institutions that adjustments be made to take the edge off instability in order to enable a sense of continuity.

Cazdyn's approach to time, as understood through life-sustaining technology and diagnostics, has been particularly useful for thinking about the lives of residents of floodplain neighborhoods in Semarang. Providing fixes to infrastructure, for example to reinstate water flow, has become a natural and remedial governmental response to increasingly frequent tidal flooding events. The language of repair pervades government plans. Dysfunction has become business as usual. Semarang's government spends large parts of its annual budget on raising and repairing streets—expenses that offer no long-term stability. Residents are still dealing with the breakdown of infrastructural systems on a daily basis, facing the spatial contradictions of drainage plans and infrastructural improvement projects. They are confronted with the "inevitable gap between what is attempted and what is accomplished" (Li 2007, 1) by development schemes.

In addition, it is often unclear who is responsible for infrastructure: the state, the local community, or private firms. Local residents often engage in relationships with a variety of available actors that in turn provide funding for repair and maintenance. Invariably, this occasions inconsistent maintenance and provision of infrastructure. While these improvised arrangements or "social projects" of fixing may be read as forms of resistance, I consider them as necessary for survival in the chronic present; they are seldom voluntary. As these new aggregate forms (of social and technological life) are exposed to various stresses, they often tend to fizzle out. An aggregate form of social and technological life is the use of neighborhood pumps along rivers in Northeast Semarang. These pump communities depend on neighborly solidarity, taxation, and other coordination efforts. In addition, maintenance relies on funds from private institutions or pro-poor initiatives. Funding cycles often don't match the temporalities of this machine. Pollution may disrupt the work of the pump at any time, and rising diesel prices may also threaten the feasibility of these projects. Instead of "making anything like a definitive event occur in the world" (Povinelli 2011, 10), projects often short-circuit; they can thus be understood as "efforts," in an individual and collective sense, to endure ongoing structural harm. Such efforts more often than not fail to disrupt the temporalities of dominant social projects yet exist as an ethnographic reality that demands our attention because it gestures to an "otherwise," and to the futures currently buried in the everlasting present. In the context of this book, I understand participatory local development projects and infrastructural modernization schemes as dominant projects. Ethnographic story-telling aims to document distinct human efforts to live with, subvert, or elude dominant forms of knowledge and power in precarious settings. This perspective illuminates the very plasticity of time and infrastructure in order to contextualize human becoming on borrowed time (Biehl and Locke 2017).

In chapters 3 and 4, I demonstrate how from within the context of economic stagnation brought about by failing infrastructures and limited resources, projects nevertheless emerge that agitate and divert (water) infrastructure. Yet while these projects, such as community-run neighborhood pumps, defiantly continue to manifest, they often create more work than ease. Following Elizabeth Povinelli, I pay particular attention to "modes of exhaustion and endurance that are ordinary, chronic rather than catastrophic, crisis-laden, and sublime" (2011, 132). The case of Kemijen demonstrates that within the confines of the chronic present, development projects follow no specific goal but mainly serve the purpose of solving problems and sustaining life at a bare minimum. Individuals have no choice but to embrace the unruly texture of everyday life; to make deals, hustle, repair, and renovate.

The Work of Crisis

What happens when a foreign actor claims to have the solution for a chronic malady? A Dutch water development project began its own restorative work on the Banger River in 2009. This "polder" building project hinged on problematizing the local supply-on-demand style of flood prevention and suggested a rethinking and remodeling of flood control. While it criticized an absence of long-term thinking and residential involvement, its mandate relied on cosubstantiating the government's imaginary of a *rob* crisis. *Rob* was seen as a signal of the apocalyptic potential of future sea level rise and helped construct the responsibility of floodplain residents in technical adaptation.

As Janet Roitman (2013) has shown in her programmatically titled book *Anti-crisis*, crisis talk is imbricated with power and authority. For example, after the 2008 financial crash, subprime trading was quickly indicted for causing ineffable suffering, thus glossing over the fact that it was a well-established practice and entirely mundane prior to the crash. Relating the subprime crisis to reckless trading obscures a systemic configuration and a mode of existence that harbors and thrives on risk-taking. Roitman argues for the analytical purchase gained by considering crisis as a secondary observation of events. Relying on Niklas Luhmann's theory of social systems, she reminds us that all social systems are self-referential and reproduce themselves by encoding their environment in a language that is specific to them. When encoding a given historical moment as crisis, a system self-consciously demarcates the present from the previous moment through autopoiesis (see Gershon 2005). Therefore, crisis can be identified and addressed in its particularity, without ever having to consider or address flaws in other social systems, such as the economy or government, or the ways in which they interact with each other. This view of crisis gets us closer to questions of power and temporality. Often, powerful figures attempt to imprint their interpretation of critical events on history in order to usher in significant change. When George W. Bush compared the post-Katrina situation to a state of exception, he discounted segregation and poverty as playing an essential role in the making of the disaster while simultaneously legitimizing the deployment of special armed forces. In the case of Semarang, I illustrate how Dutch water engineering teams working in Indonesian cities employ the logic of crisis and how it caused a novel polder authority to materialize. In the final chapter, I speak to this particular assessment of crisis and show how this evaluation comes with a specific vision of the way forward. The polder project treats recurrent problems with floods in a way that casts *rob* as a result of a crisis of management. It argues that both the local administration and residents develop ineffective, piecemeal responses to a sociotechnical problem. In a way, this assessment suggests a departure in Indonesian development politics from "rendering technical" (Li 2007) and fixing, and a turn toward nonexpert publics. However, the crisis assessment introduces a new vision of time into the lives of Kemijen's residents. Given that this vision doesn't address the "meantime," that is, it doesn't validate individual projects of endurance as responses to chronic breakdown, the project does not render social organization technical but the population's very relationship with the future. In light of this, I argue that the meantime remains a temporary location devoid of political mobility, despite the bottom-up design of the polder project.

In 2009, when construction started in Kemijen and in turn produced unprecedented attention to the subdistrict's infrastructural situation, residents were forced to inhabit new relations with the state and its temporal horizons of development. Governing schemes "empowered" local councils and encouraged participation while discriminating against idiosyncratic, that is untimely, investments in infrastructure. These fragmented municipal investments allowed communities to survive as they funneled assets into the area and repaired the technological artifacts of colonial and postcolonial infrastructuring. Residents remained rightfully skeptical about public development programs, in which they participated to varying degrees while paying minute attention to the changing arrangements of infrastructure, nature, and projects. The feeling of being "stuck" in a hardening world is a direct outcome of a state intervention that undid many residential arrangements while introducing a life that revolves around broken infrastructure in need of constant repair. "Floating," as I call it, is the result of residential efforts to synchronize with the inequitable effects of breakdown. Through the acquisition of public grants and private donations, the buoyancy of hundreds of urban communities living in Semarang's North is preserved for the time being. However, this buoyancy depends on increasing synchronization with dominant institutional rhythms and political culture. In the eyes of Semarang's government, the highest priority is to replace peoples' fragmentary interventions with water systems with a new standard of flood protection—that of streamlined polder technology. Constructing a polder is supposed to equip littoral communities with the timeless model of (democratic) Dutch water management. While Dutch water management suggested a "sociological terra nova" of flood prevention, it ended up reproducing the chronic mode of life in which floodplain residents remain threatened by flooding and have no legitimate political say in the area's future development. The polder itself extends a present in which residents must mobilize their own resources to prevent being absorbed by the swamp.

A Fractured City

Home to approximately 1.3 million people with varying ethnic backgrounds, Semarang defies easy description. The city—with its coastline of roughly fifty kilometers—looks and feels different depending on your chosen port of entrance. Arriving by sea or plane, one initially perceives a vast swamp, fields of water that stretch into barren land, fishponds, and thinned-out mangrove patches. Anchored ships wait outside the harbor, a massive and heavily securitized area where workers process goods, passengers, and raw materials 24/7. Arriving by car from the east, one passes refineries, hospitals, and university campuses before driving through a densely inhabited and regularly flooded area called Kaligawe. On Jalan Raden Patah, a dusty one-way street, one catches the whiff of fish emanating from the market Pasar Rejomulyo and hovering over abandoned factories and dormant assets. Approaching from the west, Siliwangi Road curves past the extensive airport and ducks under flyovers before leaping over the West Flood Canal, Semarang's epitome of infrastructural progress. It is the most manicured entry to the city as more and more people arrive by plane or by car from the capital city, Jakarta. Approaching from the south are students, service and factory workers, packed market vendors, government employees, and domestic tourists from other Javanese cities, such as Yogyakarta or Solo. The main road between North and South Semarang becomes a hazy mess every day, where public and private buses, motorcycles, cars, and taxis vie for space on the roads winding down the steep slopes toward downtown. These historically and geographically different points of entrance are also lines of flight—they prevent the city from entering into a single mode of existence, opening it up to various external cultural influences, land-holding regimes, and climates. These influences make life particularly difficult for entities charged with governing the use of space in Semarang. Throughout the period of my research, one of the biggest concerns of Semarang's Planning Agency (Bappeda) was illegal land conversions. Its workers were constantly trying to catch up with the changing status of land in all districts, but especially in the South, which forced them to strategically uphold or modify the city's spatial masterplan.

I experienced the urban fabric of Semarang as a confusing mix of business-oriented zones, residential areas, and wholly indeterminable spaces. Between inner-city malls, stadiums, and government buildings, one often still happens upon mixed-income neighborhoods, or *kampungs*. This scattered form of urbanism is most characteristic of the North or low-lying part of Semarang (*kota bawa*). Its counterpart, the hilly South (*kota atas*), is less promiscuous spatially speaking and more coherently zoned: space is more clearly delineated according to land use and divvied up among universities, gated communities, and transportation. Leaving and entering the city by car or bus, one glides smoothly over an immaculately paved toll road cut into ravines while passing expensive suburban developments. The South is a popular pastime area for all classes of society, as its urban development bleeds into popular coffee plantations, sparsely inhabited areas, and weekend tourism destinations. The coastal North doesn't possess the pleasant feel of fading into a nonurbanized nature. Here, the mainly failed attempt to appetize the waterfront, the drive-in beach Pantai Marina, requires visitors to pay an entrance fee—which quickly led to its loss in popularity. The shore is otherwise heavily industrialized and hardly accessible to residents.

Downtown Semarang consists of multiple bigger and smaller economic centers, but the shopping district around Simpang Lima is considered its kernel. Downtown is packed with impressive buildings, such as the Dutch-built municipality and allegedly haunted Lawang Sewu, the former administrative center of the Dutch railroad company, which was turned into a museum and event space. The economic and symbolic gravity of Semarang's city center will certainly grow given the recent completion of several private developments, such as high-end hotels and malls. There are still some informal markets where restaurateurs and private householders shop for cheap groceries, such as Pasar Peterongan, but many of these markets have disappeared as a result of urban redevelopment plans or river normalization.

Semarang seems to struggle with an existential crisis provoked by the mass killings and violent elimination of the country's political Left in the 1960s (Hadiz 2006). It was this Left that helped define Semarang, where charismatic figures like Tan Malaka built the resistance against the Dutch and imagined an independent country. As birthplace of the Indonesian communist party, the Perserikatan Kommunist di India (PKI), and home to a large and well-established Chinese population, Semarang is often nicknamed the "red" city (*kota merah*). In the years since Indonesia's independence, Semarang was associated with a "vibrant cultural, religious and political life" (Eickhoff et al. 2017, 534) on account of its multiethnic makeup, Western-style urban policies, and deeply cosmopolitan identity. It seemed to carry forward the egalitarian ideas of a comparably progressive colonial city council that "was leading the way in attacking issues of sanitation, public health, housing regulation and town planning" (Coté 2002). At the beginning of the twentieth century, Semarang attracted many progressively minded Dutch architects. As Hans Versnel and Freek Colombijn (2014, 131) noted, "Semarang, the third largest city of the Netherlands Indies at the time, had the reputation of being a pioneer in urban development, planning, public housing, and kampong improvement." It is not for me to say when and why leaders of Semarang lost this progressive edge and stopped aspiring to utopian futures, but anticommunist and anti-Chinese sentiment in Indonesia's postcolonial era certainly had to do with it. The state-facilitated mass killings of 1965 deeply rattled Semarang's progressive elite, and the genocide "continues to reverberate through everyday life" via a diffuse network of "connected and interacting narratives and practices" (Eickhoff et al. 2017, 532).

When asked what made Semarang lose its forerunner status in terms of modern urban development, some planners I spoke to explained that Semarang fell out of grace with the nation's political elite when authoritarian president Suharto took office. His Surakartan wife had no interest in the syncretic mosaic of Semarang. Today, efforts to recognize, get to know, and recycle Semarang's colonial era, a time when the city flourished economically, politically, and culturally, arguably best express a kind of post-Suharto soul searching. While such efforts romanticize Dutch architectural feats and urban planning, they also selectively memorize an unaltered Javanese identity. This retroactive relation to an authentic cultural identity can also be glimpsed in the city government's fairly recent ordinance instructing employees to wear traditional batik at work every Tuesday. In this nostalgic engagement with Semarang's Javaneseness, Semarang's achievements in city planning in communist times are effectively bracketed out.

In Semarang's North, where the city's vast floodplain fades into the Java Sea, neither colonial nor postcolonial infrastructures serve a clear purpose anymore. Patched-up riverbanks and dams, houses and roads, reflect at best a fractured design, with multiple departures from the grand plan. In view of uncertain pathways to urban coexistence with the remnants of plans gone awry, residents borrow time against the decline of their environment, as a form of risk-hedging. The *jalan santai* event related previously sums up modest but existential aspirations; in the context of recurrent exposure to toxic river water, caring for the body means caring for the neighborhood as a necessary form of exhaustion. In fact, walking events are promoted and encouraged by the subdistrict government, which rewards small grants and promises of material support to harmonious and self-caring neighborhoods.

The Swamp, the Kampung, and the Margin

Although Kemijen's kampungs are not officially slums, they are built in a "slum geology." According to Mike Davis (2007, 121), the geographical definition of a slum is a hazardous, health-threatening location. Throughout Semarang's history, the coastal swamplands have been such a hazardous location. According to Michael Taussig, despite braving the elements of a volatile environment, swamp people are barely noticed by historians.

> Much of world history includes such people, just as much of world history can be summed up as the drainage of vast wetlands and building of dykes against rivers and oceans. While a romanticized nomadism as a trope has caught the eye of many Deleuzians, the swamp people of world history seem predestined to invisibility. None of that wild movement on swift horses across the deserts and plains. Just vicious mosquitoes, sludge and wetness and kids making sandcastles on the sandy streets of riverbank villages in the dry season, awash with water and glue-like mud in the wet. (Taussig 2015, 62–63)

Similarly, the constant adaptation to changing water levels, and the labor required to keep houses, floors, and feet dry, have not been a central concern of anthropologists studying Semarang or other Indonesian coastal cities. I was confronted with this lacuna when I first began to study the former swamp. There are no photographs of Semarang's coastal kampungs from the colonial period, or perhaps they remain to be found by someone with better archival research skills. What I did infer immediately was that large parts of Semarang's swamp have been bulldozed over and filled in. As in other deltas, a kind of terrestrial infrastructure that Atsuro Morita (2017) relates to a modern Western orientation to infrastructuring coasts became the predominant mode of dealing with wetlands. But in the North of Semarang, these infrastructures also produce wetness. Today, Semarang's littoral is a place, to use the words of Lesley Stern (2017, 20), where things are not so solid and where liquid does not always flow. What, then, are the heuristics with which we appreciate human existence and world-making projects in the swamp? I suggest we consider the swamp as a marginal space—in both geographical and epistemological terms. I define it as a place in constant transformation, evolving at the behest of infrastructural landscape-making projects in which its inhabitants play diverse but rarely directive roles.

Modern water infrastructure and technologies, such as pumps and riverbanks, that suppress the swamp (Giblett 1996) as well as the sociological and economic composition of Kemijen's kampungs complicate the productivity of a comparison between the swamp and the slum. Specific aspects of kampung sociality add further complexity. While the history of Semarang's North demonstrated many characteristics of a "slum," my field site was often portrayed by residents as transcending that categorization: as a space in transition, where promising examples of development and change were fomenting. Residents' desires to transform the area were expressed in civil engagement in projects of cleaning and depolluting, relatively strong participation in community-based state programs, and aesthetic as well as economic investments in individual property. For example, many neighborhood groups organized pumping communities that operated within a traditional framework of communal caring.

The contradictions in such representations of the area have a great deal to do with *rob*. Underneath heaps of concrete and multiple layers of dirt, the swamp still lurks. Tidal flooding, regular monsoon floods, and polluted vital kampung infrastructure reinforce the swamp image that haunts Kemijen. Although *rob* indexes the constant necessity of repressing the swamp, I see it as a potent transformative agent. *Rob* signals the persistence of Kemijen's swamp geology and its imagined undesired elements—squatters and gangsters. I argue that the infrastructural labor demanded of Kemijen's residents is a kind of affective investment in the future of the kampung, one that seeks to prevent the community's relapse into morally dangerous sluggishness.[2] To illustrate this, I describe forms of state-exacted and affective urban swamp labor that, while being underexposed in the anthropology of urban Indonesia, have transformed and continue to transform Kemijen, a marginal place in social, political, and spatial terms. Existing literature on the kampung allows me to interpret this labor as a strategy of endurance and survival.

Pioneer settlers of Semarang's swamp had to put up with malaria and other epidemics while living under colonial tutelage, which effectively dispossessed them of land, barred access to resources (e.g., agriculture and water), and forced them into wage labor (see chapter 1). Today's downstream communities continue to be disproportionately exposed to health risks related to environmental pollution. Yet drainage technology, embankments, and dikes have changed the area profoundly, fashioning a concrete edge into the landscape and allowing for the expansion of a modern transportation network (roads and railway) into the former swamp. As I have argued here, it is from this nexus of materialities that *rob*—as reminder of the swamp—began to emerge. I suggest that transforming the swamp in positive ways depended on being included in the imagined community of the state through a process of "strategic misrecognition." This idea, articulated by John Bowen (1986), refers to the role that state-sanctioned, cultural forms of kampung organization and labor played in founding the Indonesian nation-state. After being officially recognized in 1949, the Indonesian state tried to extend a system of political and cultural power to all its citizens. Subordinating the kampung to the state required the inclusion of state cultural representations in intervention strategies (Bowen 1986). These interventions were often framed as *gotong-royong*, an instantiation of Indonesian state philosophy that Bowen translates as "mutual assistance" (545). Gotong-royong, as a state-sanctioned, local system of reciprocity, defines "obligations of the individual toward the community, the propriety of power, and the relation of state authority to traditional social and political structures" (545). Gotong-royong defined the relation that kampung communities entertained with the center of power. While depoliticizing the kampung by distancing it "from the rest of the state machinery" (Sullivan 1986, 85), gotong-royong endowed the kampung, the outcast of the colonial city, with a new status. Vice versa, the government expected kampung residents' "voluntary" engagement in state projects. Bowen (1986) identified a form of historical gotong-royong labor that resembled corvée—forced labor on mostly public infrastructure.[3] It was such labor that allowed and continues to allow for the effective transformation of the swamp into a "dry" space with the potential to properly become modern. But "misrecognizing" forced labor for cultural duty also helped develop a legitimate governing structure from within the swamp that was in line with the cultural norms of the state. In other words, draining the swamp and modernizing it, as an "outward conformity to state demands," was a condition of becoming a member of the "imagined community" of the nation (Anderson 1983).

Drawing on John Sullivan's (1986) analysis of affective kampung labor in Yogyakarta, I hold that infrastructural repair as well as kampung beautification efforts not only relieved the state of important expenses but also fostered a sense of community. While not being properly voluntary (or *sukarela*), this type of affective labor made the swamp into a concrete and legitimate space, through filling in and hardening out its marshy shallows. Gotong-royong created not only the cultural foundation (i.e., legitimacy) of the coastal kampung but literally its geo-ontological foundation. As Jan Newberry (2008) has further argued, kampung communities are a spatial organization of labor. In turn, specific forms of toiling in the swamp aim to (re)produce the kampung.

Today, unpaid "community" labor, often referred to as *kerja bakti*, is coordinated by various actors and collectivities that facilitate infrastructural repair, often through the cultural rhetoric of gotong-royong. The Indonesian local government administration system, which has proved to be one of the "most durable infrastructure[s] in changing Indonesia" (Newberry 2018, 199), plays an important role in mobilizing and coordinating community labor. As Martha Gay Logsdon (1974) has shown, neighborhoods are organized into two levels of administration: the smallest unit is the Rukun Tetangga (RT, harmonious neighbors), and the Rukun Warga (RW, harmonious citizens) exists above the RT.

As I demonstrate in this book, local antiflooding labor relies on the administrative infrastructure of the RT/RW system but also operates through other networks. This book's description of ecological transformation attests to organizational, institutional, and cultural changes and overlapping sovereignties within infrastructural projects. Notably, chapter 4 portrays unprecedented social relations that emerge between residents, the state, and international development agencies, namely through the recent proliferation of nongovernmental organizations (NGOs) and public–private infrastructure projects.

In addition to a governing role, it has been argued that the RT/RW system fosters "neighbourly cooperation" and a sense of community (Guinness 2009, 12–13). Drains and roads, as Sullivan (1986, 79) showed, are "foci of communal sentiment: hard evidence of neighborship. They help to confirm the existence and value of community for each member." Following Sullivan, I show how efforts to cooperatively manage drainage is a way of practicing community. In fact, Sullivan warns against a plain economic interpretation of social cohesion within neighborhood groups. He describes them as cells, or even "closely knit families," which "pool and divide labor to accomplish diverse tasks: household chores, home repairs, child minding, caring for the sick and disabled, shopping, and much else" (73). Such transactions are not merely rational responses to often "bitter need" but go hand in hand with sharing "human warmth, hopes, fears, sorrows, and joys" (75).[4] In 2001 and 2004, the RT/RW system underwent changes in the wake of political reforms that were meant to democratize local governance throughout Indonesia (Bunnell et al. 2013). The RT/RW system was thought to perpetuate social inequality and undermine development programs by allowing certain individuals to monopolize authority at the intersections of local administration and village sovereignty. For instance, Ariel's husband, Arief, was the first democratically elected head of a community group. Before he became the head of the community, a man had held the position for twenty consecutive years. Arief strongly promoted the traditional values of community work, as outlined by Sullivan, and often underlined citizens' responsibility to perform local labor.

The subdistrict government (*kelurahan*) of Kemijen further consists of diverse institutions that carry out infrastructure maintenance in the spirit of Indonesia's participatory development agenda. These institutions, too, build on borrowed on time in the name of the community. The Badan Keswadayaan Masyarakat (BKM, a neighborhood-level residential council), which operates mainly on governmental grants but also attracts private donations, has staff to oversee and implement infrastructural repair projects. In development literature, the BKM is defined as a "community self-help organization." Throughout Indonesia, BKMs are supposed to facilitate and organize community participation in development projects. The BKM involves communities in decision-making, money management, and delivering services. They often receive funds through Indonesia's Program Nasional Pemberdayaan Masyarakat Mandiri (PNPM Mandiri, National Program for Community Empowerment), a "national program that is at the heart of the Government of Indonesia's effort to reduce poverty."[5] BKMs can self-implement infrastructure projects by hiring contractors. However, a study conducted by the Asian Development Bank (2012, 32) showed that BKMs sometimes "functioned solely in the presence of development initiatives for which the participation of local government agencies was a requirement. Thus, once implementation of the initiative concerned was completed, the institution became dormant or vanished altogether." Kemijen's BKM is one of the area's most active and successful organizations, according to the *faskel* (*fasilitator kelurahan*), a state-hired coordinator who facilitates collaboration between the subdistrict administration and the BKM and implements various poverty reduction tools. In Kemijen, the *faskel* was on very good terms with the BKM and entertained amicable relationships with some of my closest informants. These relationships stemmed from a mutual dependency: while the *faskel* was the gatekeeper to sought-after development funds, his work was evaluated based on the success rate of his brokered projects in kampung communities. Patrick Guinness (2009, 239) has argued that PNPM Mandiri has contributed to fostering kampung residents' self-perception "as responsible and capable citizens of the city and equipping them to seek out partners within the wider society in realizing their dreams." In a sense, the BKM is supposed to operate outside the RT/RW system, which was judged by reformers as potentially self-serving and corrupt, and therefore unable to effectively distribute development aid. But on the ground, the BKM and RW leaders cooperate in many ways to determine development priorities and strategically invest development aid.

The state apparatus of governance is represented not only by the RT/RW system but also by the local administration (kelurahan), described in more detail in chapter 4. In addition to governmental authorities, there are a few religious authorities in Kemijen, whose residents are predominantly Muslim. The neighborhood itself has one larger mosque located on the eastern bank of the Banger, but many residents frequent a mosque catering to a bordering neighborhood. Research for this book was not designed to address the role that religious authorities play in the management of floods, but due to the ubiquitous presence of Islam in Indonesian society, I was certainly made aware of the rhythms and obligations that religious practice weaved into the quotidian.[6] I experienced the neighborhood as particularly tolerant with regard to religious practice. How, where, and what religious rituals were to be observed was a matter discussed among relatives, not a public one. While almost all my interlocutors were Muslim, they criticized radicalized forms of Islamic practice and even took administrative measures to prevent extremists from settling in the area.

While coastal communities depend heavily on infrastructure, resources, and aid provided by the state, it should be noted that other actors were and continue to be involved in transforming the area's infrastructure as well as reproducing Kemijen's slum geology. Certain private and sometimes semipublic actors have begun to shape Kemijen's ecology. For instance, the state-owned oil company Pertamina provides the bulk of local jobs and invests into neighborhoods that house the area's (formal and informal) labor supply through corporate social responsibility schemes. The NGO Mercy Corps has financed local festivals, and some state infrastructure programs ask that initial public funds are matched by private donors. These investments have also often tried to mitigate environmental impacts from industrial pollution. Overall, this set of organs played an important role in sustaining the chronic configuration of infrastructure, endurance, and eventfulness in Semarang. Yet they could also represent platforms from which other visions of the present and the future could inform practices and the materialization of the kampung.

Sinking [In]

Ethnography, Richard Wilk (2007, 442) argues, is a site where the "abstraction of the future comes into a gritty engagement with everyday experience." Immersion in everyday rhythms and the use of participant observation as an urban research method can reveal how individuals and institutions contend over the measurable human future. But how did I become attuned to local temporalities and the ways in which they structure people's everyday lives and aspirations? As with most ethnographic fieldwork, my research relied on letting observations gradually "sink in." This means doing and living through things repeatedly and adjusting one's senses and interpretations each time. The slow pace of ethnographic research allowed for the most unexpected of insights. As Jack Katz (2001) notes, anthropologists often painstakingly look for answers to research questions while constantly agonizing over time and money restraints. However, sudden breakthroughs almost never take place in ethnographic research. Rather, more often than not, a slow process of adjusting oneself to the rhythms of everyday life, language, and local culture ends up revealing to the ethnographer that the answers to their questions were always already there. I chose the metaphor of sinking to describe my ethnographic immersion for another, perhaps more obvious, reason. My interlocutors' houses were constantly in the process of sinking. Over the course of my research, I engaged in activities that become necessary when houses sink: scooping water out of living rooms, unclogging shower drains, disinfecting skin, and boiling water before consumption. I'm not at all claiming that I lived "like" my interlocutors. Yet doing some of these regular chores helped develop a pragmatic sense of the everyday challenges that fill my research participants' lives.

So I tried to sink deeper, while staying aware of discomfort and pressure, like a free diver taking note of increasing water pressure on her body with every inch that she sinks, trusting the reflexes that allow her to plummet to quite unnatural depths. James Nestor, the author of a recent book on free diving, which I came across during my fieldwork, describes the practice as follows:

> In the first 30 or so feet underwater, the lungs, full of air, buoy your body toward the surface, forcing you to paddle as you go down. You feel the pressure on your body double at 33 feet underwater. At this depth, the contracting air will shrink your lungs to half their normal size. As you keep diving, at about 40 feet, you enter a gravity-less area in the water column that freedivers call the

"doorway to the deep." Here, the ocean stops pulling you up the surface and begins pulling you down. You place your arms at your sides in a skydiver pose, relax, and effortlessly drift deeper. (Nestor 2014, 43)

Fieldwork never feels like effortless drifting. But growing confidence in social reflexes and the pull of social life—a confidence resulting from professional training and bodily conditioning (i.e., graduate studies)—eventually produce drift. After brushing up on my Indonesian and learning some key Javanese words to pep up conversations, everyday sinking felt more like the smooth drifting described by Nestor. Yet letting gravity pull you down is a risky undertaking, as you never know when you might run out of air or where the stream might lead you.

To let go of buoyancy was in fact quite difficult in the field, for both medical and cultural reasons. Walking around in sandals when the neighborhood was flooded caused an itchy rash to appear on my legs within seconds. But testing the waters was the expected thing to do. Of course, my academic future also hinged on successfully conducting "deep," engaged research. Gratefully, my friend and informant Adin often lauded my ethnographic approach to studying kampung life, contrasting it with the methods of other research teams that would never stay in the neighborhood for any length of time and often only interviewed him once.

Over the course of my research, I became amazed by the sense of buoyancy in riverside kampungs. Instead of agonizing over flooding and necessary repairs, people often found energy and time to host events that many guests attended. Festivals were put together to celebrate community projects, cheerful preparations for the National Holiday (Hari Kemerdekaan) took place every year, and men played chess well into the night. Seasonal flooding seemed to bring people closer together. Sinking wasn't visible to the eye and cannot be "felt" with the physical senses. It wasn't necessarily easy to directly correlate flooding with suffering. Through an optic focused exclusively on the everyday, the long-term bodily and social effects of sinking couldn't be fully accounted for. I needed to listen closely to peoples' life stories, trace their itineraries, and understand their life choices. While floods demand daily attention to the materiality and functionality of infrastructures from residents, other events (such as elections) and turns in life (such as migrating to the city, meeting one's partner, or leading an activist life in the Suharto era) also come to inflect the experience of flooding.

Studying the permutations of the river, infrastructural shifts, and the social experience of flooding also required an openness to serendipity and spontaneity: an ability to make decisions on the spot and step outside one's routine. Community meetings (forums for investment decisions and project development) take place regularly, but some weeks they can be scheduled at random. Sometimes, I would receive a text message just minutes before a meeting began, which sent me racing on my bike through rush-hour traffic. Alternatively, I would suddenly hear my name being shouted across the street; minutes later I would be sitting on the back of a motorcycle, heading off to some impromptu debate. In the lives of riverside dwellers, serendipity plays an important role as well. The ways in which subjects position themselves in relation to development projects, politicians, and bureaucrats demonstrated an exceptional flexibility. The resident Wahyu struggled to catch up with his many discrepant tasks: in between delivering ice cubes to clients, coaching two soccer teams, and ordering kiosk supplies, he met with local authorities, drafted and redrafted invitations to community events (*sosialisasi*), designed festival banners, and attended government meetings. This serendipity comes into being through other institutional rhythms or budgetary cycles. At times, the way these institutionally fixed timelines figured into the daily lives of the riverside dwellers produced quite unexpected encounters.

Book Architecture

Chapter 1, "Becoming: Semarang's Swamp in Late Colonial Times," introduces the reader to North Semarang with a focus on the emergence of settlements in the coastal marshland. In addition to field notes, I draw on secondary literature and colonial maps to argue that the development of Semarang's littoral was intimately connected to the pathologization of the northern wetlands. The industrialization of Semarang's shoreline and the expansion of Java's railroad system toward the end of the nineteenth century produced an increased demand for labor but often squeezed rural migrants into socioecological niches. Unwelcome in the central quarters of the Dutch and Chinese, who controlled trade and government, these migrants began to construct houses and farms in the wetlands as they provided space for agriculture. The administrative treatment of such spontaneous settlements (at that time already called kampungs) was based on cultural and racial differentiation. The result, a geographical divide between the "native" swamp and the cosmopolitan south, owes its origins to colonial rule and spatial planning and is both actualized and perpetuated in today's urban politics.

Chapter 2, "Stuck: Never-Ending River Normalization," describes more recent government interventions in the Northeast. Through the 1970s, a time of dramatic economic growth, Semarang experienced repeated waves of urbanization. Access to land and reliable sources of income became scarce, while the environment deteriorated. The government came to consider northern kampungs as problem zones made up of illegal squatters and controlled by criminals. The New Order regime aimed at controlling the kampung by means of infrastructure. River normalization—the widening and embanking of rivers—went hand in hand with evictions and displacement, similar to more directly political "normalization" efforts under Indonesia's authoritarian president, Suharto. While residents mourn the disappearance of a lush and self-reliant environment, river normalization also made the kampung more transparent and, most importantly, promised to reincorporate the area into the body of the city by virtue of its cleanliness and integration with the city's grid of modernity. The last section of chapter 2 is a speculation on the future of Semarang's floodplain: dreams of "killing" coastal rivers, damming them in the estuary, is the latest iteration of an improvement ideology that pathologizes the swamp and its residents and promotes modernization as a goal for all.

What does everyday life look and feel like in present-day North Semarang? Chapter 3, "Floating: Endurance and the 'Quasi-Events' of Living with Flooding," provides an impressionistic account of the lived quotidian by focusing on the experience of recurrent and permanent flooding. Water regularly breaches the material limits of houses, presses up against water banks until they collapse, and stagnates in kampung streets and people's houses. Residents have organized politically and socially to deal with the recurring flooding: they have acquired pumps and formed pumping associations to restore water flow, often by taking on crucial governmental tasks. As a result, *rob* has become the pulsating vein of economic and social, affective life in the area. I describe the hardship of inhabitants' lives: How many times have they leveled up the floors of their houses without knowing the exact speed of subsidence? Where does the community diesel pump really get them? I show how residents' endurance is textured by prescriptive state programs that encourage participation and self-governance. The *rob* "crisis" at least attracts new infrastructure and community development projects that allow residents to improvise and repair prevention mechanisms. But these programs never reliably restore essential infrastructure; they just patch them up. The result of the coming and going of projects is that residential coping efforts remain unseen or are framed as dilettantish interferences that do not disrupt the order of things (material and social).

Chapter 4, "Figuring: Environmental Governance and the Political Affordances of Infrastructure," speaks to the political "affordances" of river infrastructure by showing how locals develop sensibilities to hydrology as a result of shifting political circumstances. Broadly, it asks what it means to be a political subject in contemporary Semarang—a city lauded by the World Bank for its bottom-up development style and considered acutely endangered by climate change. The chapter looks at the small development projects of neighborhood groups (activist and nonactivist) and how some residents lobby for increased governmental support by portraying themselves as endangered subjects with concerns about the environment and local well-being. I describe the perks of governmental projects that often subsidize more than just infrastructural adaptation to rising sea level and often promise follow-up projects. The result of this is a constant hustling for projects at the municipal and provincial level. By describing the kind of organizing that goes on around projects, funds, and government policies in light of the Semarang government's new fixation on grassroots politics, I show how strange new assemblages emerge: for example, an alliance between former activists and the subdistrict administration. While the latter used to be a governmental tool of social control, neighborhood organizers and the administration now closely collaborate to qualify for funds through state development grants (like Penataan Lingkungan Permukiman Berbasis Komunitas [PLPBK, Community-Based Environmental Regulation of Settlement Space]) and the participatory development scheme called "Musrenbang."

These funds are mostly used to renew and repair streets and kampung infrastructure, while little money is provided for community development, education, and internal economic growth. Kampung residents thus remain the agents of state-concocted processes of progress and development. In this chapter, I suggest that infrastructure has undergone a change in meaning: stacking layer upon layer of pavement fails to afford a glimpse of a glorious future. I illustrate this through the juxtaposition of pembangunan, or development, to peninggian, the practice of elevating infrastructure. Peninggian is not identified with nation building. Rather, it suggests a meantime, a temporary location, where residents of Semarang's coastal area can "float" for the time being (*sementara*).

Chapter 5, "Promise: Remodeling Drainage," focuses on an antiflooding project initiated jointly by the Indonesian Ministry of Public Works, the municipality of Semarang, and the water authority of Rotterdam, the Netherlands. The story of this project forms the backbone of this chapter because it offers a window onto the uncertain future of Semarang's coastal neighborhoods. Despite its promise of political transformation, the Dutch–Indonesian project fosters a dependency relationship with technical solutions, stifling transformative change. I reconstruct the genealogy of the project by drawing upon participant observation, interviews, archival documents, and news articles. A main line of analysis concerns the crisis imaginary of the project. Drawing on Western scientific knowledge and "centuries of experience" with water management, the Dutch project leaders conjure up a sense of an unprecedented flood "crisis" by foreshadowing a crippled and retreating shoreline—with whole neighborhoods swallowed by *rob* in the near future. As brokers of the Dutch "polder" project, two residents try to rally support in the affected kampungs for the new antiflooding scheme. I describe their work in the project to show how their hopes for progressive change begin to falter as the promising polder turns out to be a simple technofix. While the polder ultimately delivers "dry feet," it fails to change how water management is done in Semarang. In fact, the polder is set to expire after about fifteen years of operating, in view of dramatic rates of land subsidence. The project therefore has to be considered as one step in a series of measures to manage breakdown. Residents, educated to act as enlightened and empowered citizens, thus find themselves reabsorbed into the chronic present, where crisis is managed by a largely unaccountable state and its agencies. While politicians now seem to walk away from the project claiming to have solved the flooding problem in the city's North, my interlocutors remain unconvinced. The chronic condition that they wanted to change was not the rising floodwaters but the hegemony of normalization.

Building on Borrowed Time shows that tidal flooding in the coastal neighborhoods of Semarang unfolds through various time frames: the life spans of infrastructure, the maintenance activities of residents and government interventions, and the temporality of *rob*. Suffering from tidal floods cannot be explained by pointing to climate change alone, nor to poverty or the failures of development. Rather, the many ways in which people and institutions deal with floods reveal a multiple, layered, and uneven catastrophe. I see this book as a cautionary tale: improvement schemes devised by powerful institutions produce a chronic present, a temporal location overshadowed by risk and devoid of political transformation. As my friend and informant Adin once put it, perhaps the best action would be to let this world sink into decay, that is, let the sea reclaim the swamp along with its communities. Does this acceptance of loss, of the city to the swamp, harbor the promise of a future with possibility?

1

BECOMING
SEMARANG'S SWAMP IN LATE COLONIAL TIMES

> Swamps, they used to be called, dank places where bugs multiply. As if by magic the disorder of the world will be straightened out. Rarely if ever with such writing do we get the sense of chaos moving not to order but to another form of chaos.
>
> —MICHAEL TAUSSIG, *The Corn Wolf*

WHEN I FIRST ARRIVED IN SEMARANG, the infamously flood-prone capital city of Central Java, I didn't know anyone, except for Mila. It was pouring when my taxi pulled up in front of her house, and we were both drenched by the time she managed to unchain the front gate. Once we were safely out of the downpour, we sat down on plastic chairs in Mila's living room. My gaze wandered around the room as she got up to fetch us some tea. The main room looked old but in good shape; it had retained the distinct style of a Javanese working-class house. Low ceilings and white-brushed walls framed the modestly sized room sparsely lined with wooden furniture. After refreshments, Mila showed me the other rooms, which hadn't fared quite as well. They were gloomy and smelled strongly of mold, as they hadn't been used for years. The state-owned PT Kereta Api Indonesia (PT KAI, Indonesian Railway Company) had provided Mila's family with the bungalow-style home in the 1960s. As an employee of the company, Mila's father was entitled to subsidized accommodation. After he passed away, her mother remarried and moved into her new husband's home. Mila, who started working for a Jepara-based wood export company, continued to use the house to store furniture. Initially she only stayed overnight on infrequent trips to town; however, a previous occupant soon laid claim to the plot. Mila refused to give up the house that contained so many of her childhood memories, and in hopes that evidence of occupation would weaken the man's claim, she moved in. He persisted and threatened legal action, and so due to the looming danger of having to vacate the house, Mila never bothered renovating.

I had met Mila in 2008 when I was an undergraduate student studying *bahasa Indonesia* (the Indonesian language) in Solo for three months. She was thrilled that I had returned to Indonesia, and with the goal of moving to her hometown no less. She graciously took me in while I familiarized myself with the city and scouted out research sites and suitable accommodation. During the time that I stayed with her, she wasn't occupying any of the main rooms in the house, and just one of them was rented out to a young carpenter. She had added a modest extension for herself not long ago, a simple room with a plastic roof that leaked brown drops of water during strong rains. The unstained white plaster of the walls and shiny floor tiles of her new room created a strong contrast to the rest of the house. The room also contained a modern air conditioner and a small TV. Here, Mila could enjoy the privacy and reprieve of her own space.

The floor of Mila's room was about twenty-five centimeters higher than the house's foundation. When I woke up that first morning, I immediately understood why it was raised, and why she didn't occupy the house herself. Throughout the night, it had rained so much that the clogged gutters out front were unable to absorb the surplus water. The house flooded. Brownish ankle-high water filled the guestroom and all the bedrooms except Mila's add-on. The carpenter was grimly scooping water out of his room with a salad bowl when I got up to check on the situation. At first, I watched in disbelief as he performed this Sisyphean task, then I grabbed another bowl and tried my best to help out. Mila called in her family's retired caretaker to help restore normality. The elderly man came by a few hours later to help remove the remaining flood water with a broom and a mop. The backyard of Mila's house was permanently flooded. The next day, as we hung out in the back of the house, where she baked durian pancakes for her side business, Mila pointed to the giant puddle that her backyard had become. She explained how her "garden pond," as she jokingly referred to it, had become home to some small fish. Laughing off this unsettling remark, she said that the puddle resulted from drainage backlog. She had no other explanation for it.

Mila once recounted waking up at night when she was a child: "I got out of bed and stepped into knee-deep water; sometimes it didn't even surprise me." Flooding has been an issue in the Citarum area as long as Mila can remember—a little water in the living room like on the day after my arrival couldn't unsettle her, as long as her belongings were safe. Like most residents of Semarang, Mila wasn't scared of seasonal flooding. She had become accustomed to it a long time ago. But the stagnant water in her backyard indicated a different kind of problem; it was an unsettling presence, a sign of spreading or deepening uncertainty. She seemed to find hilarity in this chaos. In the urban settlements of Semarang's North, as I would soon find out, the problem of stagnant water was far more pronounced. The ocean had been steadily encroaching on whole neighborhoods for decades, leaving whole areas submerged. Ominously, stagnant water in the Citarum area signaled the arrival of a type of floodwater that was expected to affect coastal neighborhoods.

When I arrived in Semarang, the newspapers were rife with the specter of tidal water that threatened to disrupt the city's central spaces and functions. It substantively unsettled assumptions about the geography of floodwater—where it was supposed to show up and where it was not. The governor of Central Java, Ganjar Pranowo, was evaluating ambitious plans to build a "Giant Sea Wall" or "coastal belt" (*sabuk pantai*) to protect Semarang from more extensive tidal flooding (*banjir akibat rob*).[1] To be sure, flood water regularly interrupted the lives of many of my research participants. All my interlocutors, not just Mila, had stories to tell about flooding. In a sense, flooding not only intervened in but also connected the fates of all the people I came to know throughout the course of my research. This was especially true for those living close to the city's second-largest drain, the East Flood Canal: through regular instances of overflow along the canal, they experienced flooding like a connective tissue; it brought them together as residents of a city dealing with the all-too-present possibility of flooding. "Semarang Kaline Bajir," a popular ode to the city's gatecrasher—flooding—is testimony to a city united by a common enemy. As Abidin Kusno has shown regarding flooding in North Jakarta, dealing with water problems can unite urban communities. In response to flooding, citizens come together in crisis or close ranks to hedge future risks, for instance, by collectively investing in pumps or hydraulic infrastructure. But Jakarta is also a city divided by water infrastructure, as Kusno (2018, 22) reminds us: "various programs and facilities operate like an infrastructure of difference as communities are both united by *banjir* and yet divided by their responses to it due to the variety of solutions they [have] built for themselves." For a long time, the city of Semarang could be divvied up based on the type of flooding different areas were exposed to. The arrival of *rob* in places considered not coastal (*pesisir*) was an instance of infrastructural inversion (Star 1999). This observation not only allowed me to become aware of the technologies and practices that had risen to the level of infrastructure over the years and now seemed to break down, requiring patches and upgrades to uphold the distribution of flooding, such as the giant sea wall. Inversion also pointed to a set of social and spatial relations created by various water infrastructures and that were now threatening to change or come apart. The city was built on the assumption that tidal flooding was a phenomenon of the North. It had better stay there.

In this chapter, I examine the emergence of the North as an ecological niche located on the edge of an imagined infrastructural order. The North, as fixed in maps and imagined by policy makers and residents alike, is largely the historical byproduct of colonial infrastructures of difference. Despite analyzing historical material, this chapter intersperses archival data with stories from the present. This is owed to my refusal to recognize the North as a fixed category. Rather, I hold, it is mired in an ongoing process of becoming defined by infrastructural imaginaries and practices of place-making. Evidence of Semarang's early kampungs, specifically those located in the coastal lowland, is scant. In the words of E. P. Thompson (1975, 24), "something that left so little trace in the public print of the time . . . is a matter which should be explored further." By combing through historical documents, maps, and early postcolonial scholarly work, I seek to contribute to a tentative history of Semarang's North and not a definitive account. This history, in my opinion, remains to be written. The North and its swamp were considered an aberration of history by the colonial state, an urban externality. It was viewed as an unbecoming mess of entrenched poverty and unhygienic living conditions for a modern colonial metropolis. By offering a rereading of historical material alongside contemporary stories, I wish to fill a void in the public imaginary of the North, offering a situated perspective of northern becoming that draws on stories about humans and nonhumans. Inspired by Taussig's (2015, 26) critique of agribusiness writing, I see the absence of a northern narrative as the result of an attempt of mastery over representation. Agribusiness writing, according to Taussig, is a mode of production that conceals its means of production and desperately seeks explanations to demystify and rationalize the present. In this register, writing is mere information that shuns experimentation, poetry, and humor because they blur the clear and logical lines of reasoning. In the same way that the Dutch government sought to drain the North in order to eradicate an inconvenient social reality, agribusiness writing rationalizes chaos away. Drawing inspiration from Taussig, I attempt a kind of hybrid writing that seeks to convey a "sense of chaos moving not to order but to another form of chaos" (30).

Urbanization during colonial times depended on investments in the infrastructural penetration of the floodplain. Together, technologies of drainage and the settling of dry land formed an infrastructural ensemble that was, in theory, supposed to create a safe urban environment for all inhabitants of the city. As it happened, these areas were exclusively inhabited by indigenous settlers. A historical perspective of drainage infrastructure in the North reveals "archaeologies of differential provisioning" (Anand, Gupta, and Appel 2018, 3)—not an encompassing framework for the city's many populations but a selective and interventionist logic. Framed as a useless wilderness of bog, the North was shunned by the Dutch (except for the port). Drainage efforts in the colonial era and thereafter left a fragmented landscape, a wetland ecology in ruin, for communities only partly enrolled in wage labor. While this alone does not account for the ongoing problems with tidal flooding, an archaeology of drainage infrastructure reveals the cultural and economic logics behind the fragmentation of Semarang. It illustrates that my interlocutors were "thrown into worlds not of their making, [but] into worlds already structured by infrastructures" (Anand 2017). As Nikhil Anand (2017) has argued, infrastructures perpetuate the political relationships of the times in which they were conceived. Dutch drainage infrastructure and racial policies inscribed a teleology upon the floodplain, both materially and symbolically. This grand vision of progress materialized unevenly and effectively plunged the North into a never-ending limbo of managing floodwater, a crisis composed of the dual threats of failing drainage infrastructures and the increasingly deleterious effects of climate change.

Becoming Coastal

A few days into my stay, Mila offered to clean out another room of her house so that I could move in. She anticipated that the carpenter would be leaving rather soon since his business wasn't doing very well. She explained that she would need a man in the house "for protection." While I appreciated her offer, I kindly refused. I feared the social repercussions of living with an unmarried woman in a highly religious society. I anticipated that sharing a household with Mila would cause unnecessary suspicion and perhaps disapproval among future informants. When I explained to her that I planned to move further downstream to witness tidal flooding firsthand, she grimaced. I could tell she was disappointed, but there was also a sign of disbelief in her expression. Was I sure that I wanted to live "up there"? When I insisted that the North was the right choice in light of my interest in climate change–related flooding, she gave in but recommended that she accompany me to apartment viewings. She wanted to make sure that I lived on a safe street, which, according to her, was quite a difficult thing to find in the North.

To Mila, the North was much rougher and more dangerous than her own neighborhood even though she lived by a four-lane main street that was busy day and night. Crossing it required full concentration and considerable courage. She thought that the northern area I picked was especially dangerous for outsiders who just happen to pass through. I was immediately intrigued by this representation of the North and wondered about its causes and legitimacy. Mila made me aware of another important distinction: flooding in the North was different from the kind I had just witnessed at her place in Citarum. Unlike Mila's problems with monsoon downpours that overwhelmed dysfunctional local drainage infrastructure and caused banjir (flooding), neighborhoods in the North were constantly wet. Flooding "up there" was not only a result of strong rainfall. Neighborhoods in the North were subject to daily tidal flooding events called *rob*. Her account of *rob* reduced it to tidal spillover.

As I will discuss in the following chapters, the phenomenon is far more complex. But, more importantly, she thought that all coastal neighborhoods were affected by *rob*. The picture of the North that she painted was that of an unbecoming ecological niche that faced inevitable disaster, while her own area was mainly struggling with seasonal rains that occasionally overburdened infrastructure.

Our first trip to Kemijen, a subdistrict in North Semarang, took us through Semarang's former colonial city-center (Kota Lama) and past Semarang's historic train station, Stasiun Tawang. As we continued, I suddenly glimpsed a large body of water. At first, I naively assumed that we had reached the coast and that the water was an inlet. In reality, we were still about one kilometer shy of the shore, and the body of water was in fact permanently flooded land. Stopping at a large intersection where Pengapon Street and Ronggowarsito Street meet, Mila pointed to a building on the northeast corner (Figure 1). The two-story pavilion looked strangely majestic despite showing clear signs of ruination: the ground level featured rounded doorways while the second floor had a continuous balcony that ran along all four sides of the rectangular building. The building, whose roof was visibly damaged, bordered a saltwater intrusion basin into which it seemed to sink. "What kind of building is this?" I asked Mila, who promptly informed me that it was an old Dutch office building. It belonged to the private train company Samarang–Joana Stoomtram Maatschappij (SJS), which operated cargo and passenger trains in Central Java from 1879 to 1959. During the remainder of my research, I would pass the ruin on my way home nearly every day, and it never ceased to amaze me. The persistence of this abandoned building struck me as odd. Why had it never been demolished, since it clearly served no purpose anymore? To Mila and many others, however, it had come to embody the ruinous effects of *rob* on the coastal North of Semarang. It seemed to act as a spatial marker, demarcating where the North begins while also indexing the area's historical trajectory.

This initial trip was organized so that I could meet Adin, a longtime resident of the northeastern subdistrict, who knew residents who had a room for rent. On our way to his house, driving along bumpy alleyways that seemed to become narrower with each turn, Mila anxiously asked me how I acquired his contact. I told her I had been put in touch

Figure 1. Offices of the "Steam Tram Company Samarang-Joana" on Pengapon Street, circa 1917. The building is now defunct and permanently flooded. "Kantoor van de Samarang Joana Stoomtram-Maatschappij te Semarang," presumably by Hisgen (Java). Digital Collections, Leiden University Libraries, https://digitalcollections.universiteitleiden.nl/view/item/908205. Creative Commons CC BY License 4.0.

with Adin by an employee of Semarang's Planning Agency, Bappeda. While the answer seemed to reassure her, she only fully relaxed after meeting Adin in person. Adin was a short, gray-haired man with a boyish grin. He wore jeans and a well-worn dress shirt as he greeted us outside his modest house. Along his alley, the majority of the houses stood wall-to-wall, with a few exceptions where a small path to another alleyway had been preserved despite growing housing needs. The floor of his house, like that of his neighbors, had been recently lifted to be made safe from tidal flooding. As I would soon find out, he and his wife had also just spent a significant amount of money raising the roof. After several rounds of floor adjustments, they had begun to feel cooped in. The modest, unfinished guest room contained three arm chairs and a sofa table. On the wall hung a large framed portrait of Jesus, an almost life-size depiction in pastel colors that endowed the unfinished room with a certain solemnity.

Adin immediately struck me as a peculiar man. There was none of the standard etiquette I was accustomed to from spending time with residents from other lower-class neighborhoods, or kampungs. He seemed completely unfazed by the fact that a foreigner was interested in his neighborhood and its long-standing problems with flooding. His lack of surprise was somewhat understandable in view of well-known government plans to dredge the Banger River and normalize the East Flood Canal, which promised to turn the neighborhood into a hot spot of infrastructural development and public anti-flooding measures. Adin had become a sort of ambassador for one of these projects—the Polder Banger, a closed hydrological system designed to prevent flooding. After installation, the system would be operated by residents themselves. Despite this novel arrangement, Adin wasn't completely taken up by the prospect. During one of our first meetings, he said, "The government thinks that welfare in northern kampungs will increase with physical development, but they are wrong; it does not result in empowerment. It is useless." It was his skepticism of physical development that I found particularly striking, as it effectively pointed to the area's complicated entanglement in the city's infrastructural becoming. After this very conversation, which took place under the watchful eyes of Jesus, Adin took me on a stroll to the Banger River. During our walk, Adin described the ecological destruction that industrial urbanization had wrought on the coast of Semarang. In particular, he highlighted the gradual disappearance of mangrove forests that had once flourished along the shore. They had been destroyed to make room for transportation ways, rails, and loading sites. According to Adin, mangrove forests had once acted as effective buffer zones, natural barriers to ocean water. When Adin was young, many fishponds as well as mangrove forests had still separated his neighborhood from the shore. He remembered how these "natural" structures protected their settlement from tidal highs and formed an integral part of a "nature equilibrium" (*keseimbangan alam*) now gone. Adin's extensive knowledge of previously planned but incomplete infrastructural projects played an important role in my choosing Kemijen as a primary research site.

Today, Kemijen is divided by the Banger River, a drainage canal that runs through Semarang's most densely inhabited eastern districts before flowing into the Java Sea. Every day, around fourteen thousand people (Lassa 2012) flush excrement and wastewater into the river. According to the city's drainage plan (Rencana Induk Drainase), the Banger River constitutes a drainage "subsystem," a channel that is hydrologically embedded in the catchment area of the area's major drain, called the East Flood Canal.[2] The canals were built to dry the wetland in order to construct roads. Simultaneously, the process effectively pushed the urban frontier closer to the edge of the sea. Today, the densely inhabited residential areas of Kemijen border on the premises of an oil refinery and a gas-based electricity plant.

The ambiguous Malay term for canal or river—*kali*—is used to refer to the Banger.[3] In the following, I refer to the Banger as a river, as its physiognomy prior to embanking in the 1990s resembled more a creek lined with plants and trees. I was unable to date or find out the exact origin of the river's naming, but everyone I spoke with considered the river's name intuitive—an epitome as self-evident as seasonal change. The word *banger* means stinky or rancid in Javanese; the name refers to the strong smell of the river, which contains many toxic substances, including industrial wastes and human excretions. Since the river began stagnating, the smell has significantly worsened. I learned from conversations with riverside dwellers that the river has always been a dumping site. In the absence of urban services, it served as a ready absorber of wastes, both liquid and solid. As one person put it, when the river was still flowing out to the sea, all litter would slowly drift out of sight until it was gone. For about ten years now, the river barely flows, even in the rainy season, and so waste stays well in sight, often beginning to decay right where it was disposed of. Today's acts of disposal are thus a cultural residual of the former state of the river, which afforded this kind of action. While the canal was supposed to introduce a kind of infrastructural order into the North that I describe in the next chapter, it paved the way for residents' contemporary problems with flooding and pollution.

Infrastructural Offshoots

As I mentioned previously, historical records of Semarang's early kampungs, specifically those located in the coastal lowland, are limited and scattered.[4] For example, I could not find much material on the colonial interventions that shaped early kampungs in the North. A fairly unproductive trip to Leiden, a university town in the Netherlands that is host to a large archive of colonial records, figures, and maps, led me to the conclusion that throughout colonial history, the North of Semarang was predominantly considered a problem zone and not an inhabitable space. The records of early coastal kampungs that I did manage to find all had to do with their problematic hygienic conditions and high mortality rates. These records stand to show that the Dutch didn't aim to unlock the potential of the North as future property and tax base (Bhattacharyya 2018). Rather, the Dutch considered the area a social and moral nuisance that needed to be kept at a safe distance and, at times, disciplined, if business and city life were to proceed uninterrupted.

The hilly area south of contemporary Semarang was settled in the eighth century. It belonged to the Hindu–Buddhist kingdom of Mataram, which dispatched regular convoys of merchants to Semarang. By the first half of the seventeenth century, Semarang had become an important regional trade hub with a sizable population (20,000 to 25,000). A road connecting Semarang with the center of Mataram, a center that had withdrawn far inland, was of strategic importance to Sultan Agung and maintained under great efforts (Reid 1993, 58). The shoreline had continuously advanced north by river siltation, allowing people to cultivate swampy downstream land. For this reason, coastal neighborhoods in contemporary Semarang are located about six kilometers north of what are considered Semarang's earliest settlements (Brommer et al. 1995). This slow, nature-permitting process of urbanization changed abruptly after the arrival of the Dutch. In 1678 Semarang was handed to the Vereenigde Oostindische Compagnie (VOC, Dutch East Indies Company), and in 1708 the company established a fortified European city (Kota Benteng Europeesche Buurt) on downstream land west of the Semarang River. The Dutch were not alone in the delta: indigenous, Malaysian, and Chinese settlers had already been living on the marshy land for centuries. Until the nineteenth century, the area east of Semarang River, where I conducted the bulk of my research, remained largely uninhabited and was used for aquaculture. Throughout the nineteenth century and into the first half of the twentieth century, many of Semarang's kampungs were quite literally spaces on the side of the road. They provided shelter to the indigenous population in the employ of Dutch companies, such as the SJS, and government offices. Freek Colombijn (2002, 610) noted that before the completion of the Great Post Road (Grote Postweg) in 1811, Semarang's urban functions were concentrated along the Semarang River. Many of these functions were subsequently moved to the post road or to parallel and side roads. Developing a sophisticated network of roads was of utmost importance to the colonial government, as early maps projecting the urban expansion of Semarang show. The Dutch were invested in connecting the low-lying delta, a hot and humid zone where trading, shipping, and some production work took center stage, with the cooler and romanticized highland as well as the crop-producing hinterland. They further intended to connect Semarang with other coastal settlements, such as Demak, Kudus, and Pekalongan.

By the end of the 1910s, Dutch maps make mention of kampungs that had formed on the eastern entrance of Semarang, between the East Flood Canal and a river that possibly was already called Kali (River) Banger.[5] When Semarang's port businesses started attracting Indonesians from the rural inland and other parts of Java, the living spaces of this indigenous workforce became increasingly crowded. Over a short period of time, Semarang's "Indonesian population was augmented daily by hundreds of workers . . . to work on its wharves and factories" (Coté 2002, 320). Gradually, the Indonesian workforce populated vacant space outside the city's "primary" network of streets and rivers (Pratiwo 2004). Here, indigenous dwellers often cultivated land to generate additional income, probably taking advantage of the downstream drains that the Dutch had constructed, because they mitigated flooding and helped irrigate rice fields. The map provided here (Figure 2), which dates from 1917, shows Semarang's northeastern swampland, which contained fishponds and stretched from shoreline to the railway. On some maps, residences with adjacent farming fields are recognizable along Pengapon Street, which connected Semarang with Demak. A later map, from 1918, by the Dutch architect H. Maclaine Pont indicates these roadside residences as "kampungs at the eastern city entrance." Ideally, these unwieldy *desas* or "native" villages lying on the edge of the city would one day be integrated into a functionalist network of roads, corridors, and other productive spaces of the colonial city. Or as Rudolf Mrázek (2002, 67)

Figure 2. Map of Semarang by J.H. De Bussy, 1917. The map shows how vast the area used for aquaculture was in the Northeast. On the periphery of the city center, a number of settlements hugging drainage canals can already be glimpsed. Digital Collections, Leiden University Libraries, http://hdl.handle.net/1887.1/item:2011233. Creative Commons CC BY License 4.0.

described the colonial making of Indonesian cities, "there was a copycat order on one side; on the other side, the disorderly urban rest was to be swept away or under."

Northeastern kampungs, such as Kemijen, were roadside offshoots that existed in the margins of water infrastructure. The marshy delta was crisscrossed by rivers and canals that were part of the expanding Dutch drainage and canal system. The proximity to sewer and drainage infrastructures was a trade-off: while it allowed for irrigation of paddies and other farming practices to supplement their families' income, it exposed kampung residents to waterborne diseases and regular flooding. Tellingly, at a public meeting of the city council held in 1922, a "native" council member compared living in a city kampung to living in "backwater" (a part of the river where the water is stagnant). This environmental allegory referred to an economic and political reality: urban expansion and privatization of previously indigenously owned land had left kampung dwellers with "no property, authority, or autonomy" (Cobban 1988, 280). In view of the "disorderly urban [and dispossessed] rest" piling up in the kampungs and calls for intervention from indigenous representatives, the Dutch government decided that money had to be found for improvement, either by relocating residents or sanitizing, that is, draining the swamp and building roads into it. As Joost Coté (2002, 322) has argued, many members of Semarang's colonial society had developed sensibilities to the incongruences of "colonial modernity" and promoted an "enlightened" style of urban governance. The entrepreneurial and reformist spirit of Dutch residents new to the colony increasingly found expression in grand visions for the future of Javanese cities, articulated and illustrated in maps, housing schemes, and exhibits. Here, economic and social evolutionary theories most often provided the justification for a "dramatic intensification of colonial intervention into native life" (323). During the last thirty-five years of Dutch rule in Indonesia, the treatment of kampungs morphed into the defining object of municipal authority. Administrators of colonial urban centers, such as Semarang, worried increasingly about the "squalid living conditions of the indigenous people within the cities" (Cobban 1974, 407). How were municipalities to deal with the visibly uneven distribution of prosperity, or the "juxtaposition . . . of contiguous areas varying in physical attractiveness, population densities, hygienic conditions and standards of living" (403)? Public repugnance at the destitution of the indigenous population and fear of epidemics called for more control and stronger state involvement in the fate of urban "native" subjects. The new tangibility of starkly uneven living conditions in the colony's cities fueled transcontinental debates over the ethical responsibilities of the colonizer in view of the irreparable damage it inflicted on "weaker" civilizations (and sometimes the pauperized European population of colonial cities). As such, the uneven distribution of life and death in colonial cities produced "intensive colonial introspection" (Coté 2002, 325) and called into existence a lively discourse about the extent and techniques of colonial rule in cities. Decentralization of the colonial state apparatus had led to the creation of municipalities, which further opened spaces for reflection on the "new conception of the duties of the state" (327). The principle of indirect rule and minimal intervention that characterized the "liberal period" was considered outdated. For progressives, it was European identity as bearers of Enlightenment and progress that was at stake in educating the colonized population and ensuring their welfare. The Indies were thus always more than a site of exploitation; they were also a "laboratory of modernity" (Stoler 1995, 4, drawing on Rabinow 1989). In fact, as a response to epidemics that affected Indonesian cities, such as Deli (1905), Malang (1911), and, five years later, Semarang (1916), inspections in poor kampungs became a regularity. For example, indigenous households were obliged to bring their furniture outside "and the whole house was made clean" (Cobban 1988, 276).[6]

The *Kampung Question*

Semarang is often considered an example of progressive colonial policy, departing from the general trend of Indonesian city planning. Scholars have pointed to important innovations in urban planning that sprung from a colonial concern for "native" well-being. Indeed, toward the end of colonial occupation of Indonesia, even some Dutch related the squalor of indigenous areas to misguided colonial politics. They therefore tried to define governing problems and formulate improvement strategies. In combination, the question of how to address the problem of the impoverished native underclass and in whose responsibility such acts lay morphed into the so-called *kampung question*.

Politically engaged residents of Semarang, like the apothecary and businessman H. F. Tillema or the famous Dutch architect Thomas Karsten, played a key role in framing the kampung question. Their publications and work have already received intensive scholarly scrutiny (Coté 2002; Mrázek 2002; Nas 2003). Suffice it to say that both Tillema and Karsten argued for a gradual and planned integration of indigenous settlements into the body of the city (Coté 2014). Although the option of "slum clearing" was never off the table, it was generally judged as too expensive, hence unrealistic for cash-strapped municipalities. Visions for moving the coastal population to the sparsely inhabited highland, first brought up by the wealthy Dutch lawyer Conrad Theodor van Deventer (Wertheim and Siauw Giap 1962, 233) in 1906, were entertained for a long time but eventually dropped. Instead, modernity was to be introduced gradually into the outlying indigenous areas via sewage and water infrastructure. The Dutch architect J. J. Rückert noted that elementary steps toward improvement would require new "housing statistics, proper maps, legislation [and] financial possibilities" (qtd. in Roosmalen 2008, 297). Rückert also warned that intensive maintenance would be required after improvements had been carried out. Instead of intervening quickly, Semarang's city council had envisioned a future in which it would become intimately involved in the private lives of kampung dwellers—"during all phases of activity, administrators and planners should engage themselves in direct and close contact with the population" (297).

Tillema, who called himself an "engineer of health and hygienist," was perhaps *the* person to draw attention to the problem of water (Mrázek 2002, 56–57). He dreamed of a sanitized and regulated colony that would improve the lives of European and "native" residents. In his books, he documented in detail practices of native hygiene as well as colonial water regulation "in design and action." In Tillema's first publication, *Rioliana* (from Dutch *riolering*; English: sewerage), published in 1911, he pleaded for the introduction of a general sewer system. He argued that "as well as providing fresh water, the key to urban public health was the adequate evacuation of sewerage and drainage water" (qtd. in Coté 2002, 331). However, according to him, the districts of the Javanese population were unsuited for the construction of such infrastructure. Therefore, "adequate urban planning was essential and had to be based on hygiene considerations" (331). Tillema was a strong proponent of state-induced social development rooted in rigorous scientific practice. Despite voices such as these, which highlighted the ethical responsibilities of the colonial state, the interventions into kampungs that followed were fragmentary at best.

In fact, as Michelle Kooy and Karen Bakker's study of drinking water provision in Jakarta exemplifies, technological improvements of urban infrastructures were exclusive to the Dutch population. Drinking water provision is one example: "in contrast to the surface water from the city's *kalis* (canals) upon which natives relied, European households now had access to scientifically monitored groundwater, circulated through iron pipes rather than the corporeal networks of ambulatory water vendors" (Kooy and Bakker 2008, 379). The differential distribution of access to potable water not only created highly uneven landscapes, separated by the newly designed artesian system, and levels of regional integration, but also translated into distinct practices of hygiene and water consumption.

I should mention that while the kampung question and the welfare of the "native" population remained a major preoccupation of Semarang's city until the Japanese invasion of the archipelago in 1942, Semarang was also the home of many architectural and planning projects explicitly "modern" in character. The renowned Dutch architect Hendrik Petrus Berlage called Semarang the "loveliest town of Java" for its great examples of modernist thought and form (qtd. in Mrázek 2002, 64). In 1914 the Colonial Exhibition (Koloniale Tentoonstelling) took place in Semarang. This international trade exhibition was a site for the dissemination of modernity's primary theme, progress (Coté 2006). As "hallmarks of European cultural production" (Stoler 1995, 15), the colony allowed the world to glimpse the future of all cities, especially in contrast with the low-class, indigenous quarters. Coastal kampungs were thus a stain on the project of modernity, while equally justifying continuous colonization "work." As Mrázek (2002, 57) pointed out, "the water ordering, in dreams, plans, and actuality, in the Indies . . . worked" because the dirty world of "native" kampungs could always be contrasted to European water sophistication.

The Dangerous Swamp

Between 1900 and 1925, Semarang grew by 100 percent, which further stoked Dutch apprehensions. How was the city to absorb and govern this swelling population, especially a population that the Dutch considered unfamiliar with modern hygienic norms? The documented excessive use of chemicals around wells is a potent example of Dutch anxiety surrounding uncontrollable diseases. However, anxiety concerning the indigenous kampung was not only related to the containment of fluids and germs. In view of the collectivization of the 1920s and 1930s (Wertheim and Siauw Giap 1962), such as the creation of budding trade unions in cities (Ingleson 2001), the kampung was also conceived of as subversive body politic. Again, Rückert warned the city council that letting the situation of the kampungs worsen would eventually force the colonial government to intervene more violently in order to prevent unrest. He claimed to understand why indigenous leaders were "not at all enthusiastic about the kampung improvements" (qtd. in Roosmalen 2008, 300). In 1928 he made a plea before the People's Council (Volksraad) for a government subsidy for municipal kampung improvement works. He urged the municipality to intervene as quickly as possible, as there was no "better thinkable means of propaganda" than letting the situation further deteriorate. He warned the council that "up until now the kampong population has remained quiet but the source of unrest, the perpetual threat to peace and order remains in the so badly neglected kampongs" (qtd. in Versnel and Colombijn 2014, 134). In fact, on a map from 1913 used by Tillema to illustrate the sanitary problems of Semarang, the area known today as Kemijen shows a disproportionately high rate of annual mortality.[7]

Today, residents of the North still bear the brunt of environmental decline. Inhabiting a degraded environment not only exposes people to waterborne diseases but also demands their constant attention and energy. Especially in poorer kampungs, largely invisible to the state and public, significant unaccounted labor is required from residents to keep waters at bay and flowing orderly. I was told by residents that hypertension (abnormally high blood pressure related to high levels of stress) is one of the most frequent causes of strokes in North Semarang. It is the leading cause of mortality among Indonesians above five years of age, accounting for 15.4 percent of all deaths, according to Yohanna Kusuma et al. (2009). In Kemijen, strokes were a very common affliction among adults.

Clearly, colonial urban planning was motivated by what Michel Foucault (2003) considered the biopolitical turn in governance, where the state's "strategies of rule" (Stoler 1995, 1) revolve around the protection of life (Cooper 2008). But as Ann Laura Stoler (1995, 12) has pointed out, biopolitical concerns in the Indies articulated with the politics of race. She showed that an "implicit racial grammar" backed the sexual regimes of European bourgeois culture. This racial grammar also informed the grids of intelligibility employed by town planners in the Indies. For example, the swamp, placed particularly low in the hierarchy of livable places, was strongly associated with a degraded human needing (hygienic) discipline. Colonial representations of the urban kampung therefore posit this space and its inhabitants as a source of danger. Its inhabitants are also framed as victims of colonial modernity and capitalism, but always only as the counterpoint to a desired urban future. The inhabitants of kampungs in North Semarang were considered backward and lagging behind the trend of human evolution in the eyes of the colonial elite. While Indies plantations were controlled with brutal violence (see Vickers 2005, 17) to increase crop production, a surging indigenous exodus from the countryside to Javanese cities made it easier to simply ignore the problems of the swamp.

To this day, the shore remains an urban frontier in which poor people have to make do while facing legal and economic uncertainties. A distant relative of Adin, Pak Rianto, has lived in Depo Indah (beautiful depot), a northern segment of Kemijen, since 1975. Depo Indah draws its name from a local discontinued train depot and reloading site formerly run by the state-owned railroad company (PT KAI). Water infrastructure in the neighborhood, such as drains and canals, is dated and the streets are badly maintained. Compared to South Kemijen, houses look slightly rugged and unfinished. While there are a few recently built houses, most constructions are old and in need of renovation. When he arrived in Semarang, Pak Rianto began working for the PT KAI. He manually changed train lanes and reloaded cargo. Allegedly, he came up with the place-name after moving his family into a house he built himself on the company's premises. They were not the only informal settlers; other migrants sought out the area and built homes. While the company tolerated undocumented housing in proximity to its facilities, the fact that residents were unable to officially report to a governmental agency was a political no-go in Suharto's New Order. In the 1990s, the area was therefore annexed by the subdistrict of Kemijen.

Today, Pak Rianto, whose leanness is only surpassed by his contagious, almost toothless smile, has long retired from PT KAI. He and his wife, Ibu Ita, now live off his modest pension, augmented by some irregular income from a small-scale video and audio equipment company that they run. There are still some fishponds left in Depo Indah, a century-old trait of the North. Pak Rianto remembers that the area was quite rich in fishponds when he started his job. After his longtime employer, PT KAI, shut down its main offices and removed most of its facilities, the area became abandoned to a new threat: tidal flooding. It marks both the end of an expansive trend and the beginning of new coastal infrastructuring.

As recent developments have made clear, the neighborhood never fully settled into legality and remains illegitimate in the eyes of the government. When I met Pak Rianto, he was swept up in a legal battle. PT KAI's plans to lease its land to the government in order to construct a retention basin had forced Pak Rianto to request formal recognition of land ownership. According to the government, the land was never rezoned as residential, whereas Pak Rianto argued that land ownership was officially granted to settlers by Indonesia's "number one," President Suharto himself. Records of this transferal are, however, difficult to ascertain, and Pak Rianto's only set of proof is a worn newspaper article and a letter from the central government to Semarang's municipality. The absence of more actionable proof again shows the area's existence on the margin of the city's frame of legality, with fuzzy geological and infrastructural boundaries. One of Pak Rianto's neighbors, Pak Rozi, was indignant when students from Semarang who visited the area for a research assignment called him an "illegal" resident: "I'm not illegal! I pay for electricity and I pay taxes! Why do people call this land illegal?" Pak Rozi was even more enraged by the fact that people from outside considered the people of Depo Indah unimportant (*remeh*). They were infantilized and considered uneducated (*kepandaian ilmu kecil*).

Today, few residents still own fishponds in the area. If they actually do, they rarely if ever engage in fishing. Rather, as Pak Rianto informed me, people "from far away" come here to catch fish and tend to the ponds. Fishing no longer provides enough income to survive on, and residents therefore traded off their fishing rights. The land that Pak Rianto still owns in North Kemijen is a remnant of Semarang's swamp, a semilegal borderland between officialness and illegality. It is neither industrial nor postindustrial, neither residential nor agricultural. It inhabits an impossible location between a brighter future promised by new drainage infrastructure and current economic stagnation.

Marginalized North

Describing the "becoming" of northern kampungs in the wake of rapid urban growth before independence shows that northern communities often inhabited zones that exposed them to unfavorable and health-endangering environmental conditions. Specifically, the beginning of the twentieth century sees the emergence of urban peripheries filling up with the human "waste" (Bauman [2003] 2011) that ethnic segregation and capitalist exploitation produced throughout the archipelago. The segregationist character of cities in the Dutch East Indies forced migrants to settle within the confines of rapidly growing kampungs that were often located on the outskirts of the city and between flood-prone drainage channels. Small-scale agriculture, fishing, and aqua-farming on ambiguous land once allowed indigenous settlers to eke out a living, but the growing labor demand of factories and businesses led to the increasing commodification of indigenous labor. Gradual loss of access to land as a supplementary source of livelihood as well as the densification of kampungs exacerbated hardship and had devastating impacts on the living conditions of its inhabitants. By the twentieth century, northern kampungs had turned into surplus labor ghettos as the city's economy was not able to absorb the new floating population of uprooted peasants. Arguably, the relegation of this excess population to the margins of the city also created a growing sense of community despite ethnic diversity. Within the confines of kampungs, communities of mutual care with their own social and ethnic hierarchies developed.

Making the indigenous population of Semarang's North "even with the times," as a Dutch study argued for, necessitated the elimination of "the slum-dwellings situated in the Northern, marshy district of

Semarang" (Wertheim and Siauw Giap 1962, 232–33). The stated goal was to destroy ties to an epidemics-ridden land. The Dutch considered marshland kampungs as "neither wholly on the pre-capitalist nor wholly on the capitalist side" (230) but somewhere *in between*. Relocation was therefore considered one remedy to indigenous misery in coastal settlements and a strategy to "enframe" (Mitchell 2002, 14) subjects in industrial labor. As it turned out, neither municipalities nor the central government could find the means to achieve this feat. In particular, the central government was slow to come up with solutions.

Evidence gathered in this chapter suggests that the goal of profoundly changing the lives of coastal dwellers by lifting them up out of the swamp and resettling them was never achieved. Instead, what the Dutch treatment of Semarang's North illustrates is a form of governance that not only differentiates between populations based on geographical position but also bases its strategic disinvestments in populations by way of a racial hierarchy. This racial hierarchy—which was never straightforward in the Dutch East Indies (Vickers 2005, 25)—in turn became inscribed in urban space. In the following chapter, I illustrate the ways in which this spatial hierarchy still has a huge influence over contemporary images and ideas of Semarang. According to Pratiwo (2004, 6), the colonial period laid the groundwork for a "spatial fragmentation" of the city, which persists today. Similarly, James Cobban (1974) has argued that political and economic marginalization postindependence can be traced back to a set of colonial discourses and governance decisions. He contends that the difficulties of kampung improvement and urban infrastructure development in colonial Semarang show that "many of the problems present in Semarang today are of long standing: they were exacerbated but not caused by the rapid increase in the size of the population of Semarang after the end of the Second World War" (Cobban 1988, 269).

Vanished

I ultimately found accommodation in Karang Asem, an area bordering on the southern end of Kemijen. When I first walked into central Kemijen, I was pleased by how close it was; I only had to cross the on-ramp of a flyover to get there. I often wondered about the origins of the name of Karang Asem. For a long time, nobody seemed to know why it was called "bitter place." My downstream neighbor Arief accidentally answered the question one day while we were discussing the unused plot next to his house. He explained that his in-laws' house once stood there before they were forced to tear it down—after his in-laws passed, the sunken property had turned into a mosquito breeding ground (*sasaran nyamuk*). While waiting for a good offer from a potential buyer, they decided to plant some trees and a garden on the property. Arief showed me the plants that he had placed there and in front of his house, naming them all. He added that they all could be used for cooking or as medicine. At the end of the garden tour, he showed me a tamarind tree that people call *pohon asem*. Apparently, it was the last one left in his neighborhood. I found it beautiful; it had delicate leaves that looked like dragonfly wings as they vibrated in the wind. Arief explained that the streets of the neighborhood used to be rife with these trees, a halophyte that tolerates high levels of soil salinity. And so I learned that my neighborhood drew its name from tamarind trees that could tolerate poor sandy soils. The tamarind offered welcome shade but also carried a bitter fruit that people used as flavoring in many dishes. Just like the mangroves whose disappearance Adin lamented, these beautiful trees largely vanished in the wake of urbanization.

In the summer of 2015, I attended a solemn housewarming ceremony organized by Indah and her family. Indah, a recent economic management graduate, also grew up near the Banger River, in a subdistrict called Rejosari. We met at city hall, where she worked as secretary of SIMA, a new water authority I discuss in detail in the final chapter. When I arrived at the housewarming party, men were crammed into an unfurnished guest room reciting Islamic prayers. The small house was located in an alley off the inspection road of the East Flood Canal, just two minutes north of Mila's old place. In recognition of Hindu custom, a small offering had been placed by the house entrance to cleanse the building and keep bad spirits out, blessing the new inhabitants' relationship with the house and its immediate surroundings. Indah went to help her mother, who was busy serving refreshments to other attendants: relatives, friends from work, and new neighbors. I sat down outside to chat with Adin, who, as Indah's colleague, was here to pay his respects. On the morning of the festive inauguration, I had visited Indah's old home. The house was located in a lower middle-class neighborhood along the Banger River. While this area was not coastal, the family had been regularly affected by *rob*. Indah once showed me a video of a big bubbling puddle in front of her house that magically grew in the afternoon hours as if it were sucking water from the ground. Indah's old home, which I briefly visited prior to the move, stood out from the row of neighboring houses, some of which were miniature neoclassical palaces. She herself had grown up in this home. Upon entering, I could tell that it urgently required repair. Gladly, the new house was in much better shape. Although it lacked sunlight, as it was tucked behind a multistory building at the end of a cul-de-sac, it was a concrete construction and possessed a solid roof. The new home clearly symbolized a step up for the family, one they couldn't afford in their old neighborhood. Finding affordable property had not been an easy task, and the move was risky in terms of flooding. They had edged a little closer to the East Flood Canal, an area notorious for seasonal flash floods. In fact, in early 2016, after I had left Semarang, Indah sent me snapshots of the street by her house flooded in knee-deep river water. Then she confirmed a hunch: some of the water wasn't disappearing after rainfall ceased. It stagnated in the streets like *rob*. Indah's story of moving from one flood-prone house to another tragically shows that the hopes and strategies of poor families can be undone by the city's water infrastructure, which often distributes risk and exposure in unpredictable ways. Furthermore, it suggests that *rob* is spreading in ways that disproportionately affect poorer families.

When I returned to Semarang in 2018, Indah was still unmarried and lived with her parents. She supported them with money that she earned as secretary and youngest member of SIMA. A permanent job in the government had not come forth as she had hoped. At the time, her own family's house was undergoing renovations in response to flooding. I could tell that Indah was in a dark head space; with the job prospect gone, the idea of recurrent costly house renovations seemed daunting. Something that she had worked hard toward was slipping through her fingers.

2

STUCK
NEVER-ENDING RIVER NORMALIZATION

> There is a gap between what we provide and what people really need. The thing is we actually don't know what they really need.
> —Semarang government official and urban planner, 2015

ONE DAY, AS I SET OUT ON MY REGULAR MORNING WALK, I noticed my landlord's bird cage lying sprung open and empty on the porch of his beautiful wood-adorned house. Strong winds had unhooked the cage during the night, and Eko's precious songbird had been whisked away by stray cats. Just as Eko's daughter was finishing telling me what had happened, he returned carrying something green in his hand. He had found the bird, whose now lifeless feathers looked pretty roughed up. Eko stared at the silent songbird. "The cats killed it," he told us soberly. Then he lifted the empty cage with his free hand, walked toward the calm river, and promptly tossed the cage in the river. Without ceremony, the dead bird followed. I tried to look unaffected despite being dazzled by this aggressive, grief-laden act. As if to legitimize his act, he mumbled, "The people at the bridge will fish it out," as he disappeared into his house, letting out a brief angry grunt on the doorstep.[1]

About a year after witnessing the demise of my landlord's songbird, I attended a late-night meeting in Kemijen's subdistrict office, located about three hundred meters northwest of his home. I had expected a small, informal organizational meeting, but the *balai* (hall) was packed with residents and the atmosphere exuded the rigidity of officiousness. The residents sat on neat rows of green plastic chairs, some of them looking around for familiar faces or chatting among one another. Others were just eating silently from the snack boxes they received upon signing in. A singular wooden table had been placed center stage. From here, a projector displayed a resident-produced short film on *rob*. The *lurah*, the head of the subdistrict, had driven down from his upstream residence with his wife. At the beginning of his address, he explained that the mayor would visit their neighborhood ahead of schedule and stressed this wasn't cause for concern, as they still had enough time to prepare for the important event. They would of course arrange the usual river-cleaning event (*resik-resik kali*) to improve the visual appeal of a good stretch of the riverbank. He then sketched out the itinerary of the cleaning procession. He ended his speech by reminding them that they should send a signal to the mayor that there were people in this neighborhood who cared deeply about it. Although some residents were still "unaware" and frequently offset their efforts to maintain a clean river, he was convinced that those in attendance "all already care [*peduli*]."

The lurah's pitch was followed by another speech. The representative of the Lembaga Pemberdayaan Masyarakat Kelurahan (LPMK, Neighborhood Community Empowerment Board), who sat next to the lurah, gave a passionate address about their long-standing struggle to build a "great" Kemijen.[2] The LPMK's neighborhood suffered the most (*paling menderita*) because it was at the receiving end of the stream—where waste accumulated and the water stagnated. The neighborhood was also located where the river met the ocean and therefore experienced regular flooding. Despite this, Kemijen always seemed to be the government's most left-behind target (*sasaran paling tertinggal*). The recent development efforts, however, especially the community-based environmental regulation efforts, represented indisputable proof that the LPMK now had a clear plan (*gambaran jelas*). According to this plan, it would move the subdistrict "forward" (*memajukan*). The representative summoned the support of the residents: "let us unite [*mohon bersatu*]!" United, so it seemed, they would be able to bring improvement to Kemijen. Before ending his speech, he urged the attendants to see the mayor's visit as an opportunity to ask that "normalization" (*normalisasi*) activities be promptly carried out.

These two vignettes demonstrate two apparently contradictory views and relations with time. The first story shows the Banger River as a dump. Eko's relationship to the river enacts a representation of the North as a failed ecology: a place stuck in its ways, forever polluted and poor despite repeated efforts at improvement. In a sense, the river hails Eko—letting him enact a well-rehearsed vision of the area and invoke a specific resident-subject that inhabits this milieu: the people downstream who live off waste will even fish out a useless cage. In fact, many locals continue to use the river as a dump, in more or less visible ways. In the absence of authorities that enforced pollution laws, it happened quite often that people disposed their waste into the river. The boys and men smoking on the riverbank threw their cigarette stubs into the river; women dumped food waste and dishwater into the Banger. In addition, even when people abstained from throwing waste into the river, preferring to use designated garbage disposal sites, all inhabitants—including respectable civic and neighborhood leaders—channeled their wastewater into the river and thereby contributed to its pollution. Eko's act of throwing away a sizeable bird cage reaffirmed this relation with the river.

On the contrary, in the second story, residents are called upon as "already caring" and inclined to protect and clean the river. Here, the river's current state appears as a stain on the reputation of the subdistrict; it is perceived as visible and odorous evidence of both lagging behind the general trend of development and having been abandoned by the state. The waste-clogged river reflected badly on the area's kampungs and their constituents. As the subject of official visits, the river was an indicator of development—both moral and economic. Almost paradoxically, the river needed to be cleaned or made presentable for the area to be eligible for funding to cure periodic flooding. In order to qualify as meeting development targets, residents had to prove that they were already invested in environmental protection.

In both vignettes, the river emerges as an interlocutor or companion, a political and cultural reference point, and a site of local place making. These stories portray meaningful interactions with the river that also indicate specific orientations to time. My landlord's actions seemed to say "this is how it is," while the meeting represented a resounding interest in, and reliance on, the river as a development potential for the neighborhood, on "what could be." Like my landlord's act of disposal, the speeches invoke a subject: a subject with potentiality, if residents can unite and unleash it. This subject is the subject of normalization; it is a subject that has internalized the function and meanings of drainage infrastructure. Furthermore, it is a subject that "sees" pollution and acts upon the degradation of the environment. Importantly, both vignettes focused on the riverbank as a sort of vantage point of the present: through the riverbank, people perceive, act on, and control the state of their living environment.

Lisa Björkman has pointed to the important relational work that infrastructures do. She argues that their embeddedness in everyday life tends to obscure how infrastructures are themselves relations among things. In her ethnography of Mumbai, the disruption of water flow reveals critical nodes of urban infrastructure. According to Björkman (2015, 12), these interruptions can "work as [a] methodological entryway to the sociopolitical and material forces underpinning otherwise taken-for-granted urban processes and geographies—a means by which to explore the technologies, materialities, and politics that infuse everyday life in the city." In Kemijen, it is moments of overflow and seepage that reveal relationships between the everyday struggle of inhabiting a risky ecology and wider urban politics. Banjir, or overflow, is a semantically rich concept in Indonesia. *Banjir air mata* connotes excessive crying, while *banjir pengungsi* refers to floods of refugees. In most cases, then, banjir refers to undesirable excess (*berlebihan*). North Semarang, as a whole, has long been considered as producing excess, as I showed in the previous chapter: in terms of water as well as population, waste, and crime. It is threatened by tidal flooding (Marfai et al. 2008), but it is also seen as a threat to the rest of the city. Infrastructure in Semarang never just controls floods or reduces pollution; it implants specific worldviews in space and tries to infuse everyday life with top-down temporalities. By espousing these worldviews, I argue, residents forge new links with the government through which sentiments of pride and investment can be signaled.

While the area is a product of state neglect and weak governance (Das and Poole 2004), state control is not absent from the North. Its disconnection (Ferguson 1999) is the result of differentiation: cultural, spatial, and administrative (Tsing 2005, 42). A young urban planner working at Semarang's municipal planning agency, Bappeda, often complained to me that he didn't know how to help the poor populations of

Semarang's northern kampungs. In his view, the disconnect between the government and these places was too vast. I heard multiple such complaints about the complexity of planning problems in the North. They often sounded to me as if the government felt alienated or at least deeply confounded by these places. This epistemic divide, partly a heritage of colonial rule (Tsing 1993, 26), is reproduced by contemporary urban and development politics.

Much has been written about the Indonesian kampung, or urban neighborhood. To be clear, I neither want to do away with or append existing definitions of the kampung, nor do I want to add specific nuance to the ethnography of kampung life by offering an exhaustive description of its forms and conditions, as others have done (Siegel 1993; Newberry 2006; Guinness 2009; Simone 2010). I am more concerned with trying to understand, via water governance, the impact of political change and a shifting urban ecology on the becoming of poor neighborhoods of Semarang's North. What further motivated me to write this chapter was a desire to critique the discriminatory treatment of riverside dwellers by successive governments. While renewed governmental attention to the area signals a turn to more inclusive policies, the logic of these programs is still very influenced by the imagined necessity of controlling and remaking riverside kampungs to be more in line with elite visions of the city's future and more legible, to use James Scott's (1998) famous term. The marginalization of Kemijen, I argue, is the corollary of a planning rationality that doesn't recognize its own biases and shortcomings but considers underdevelopment and (partial) implementation of sewer and drainage infrastructure as a result of lacking residential commitment.

Fear and Danger in North Semarang

As I had come to Semarang to study flooding, I followed the course of the Banger River on a daily basis. On foot or by bike, I looked for signs or events of dysfunction and observed or participated in activities on the banks of the river. I came to know the residents of the various neighborhoods connected by the Banger. Ironically, people living downstream contended that the most dangerous people lived even further downstream—or north. Such accusations, which drew a rather diffuse image of the area as marked by violence and unrefined manners, were nevertheless undergoing a shift. In that shift, negative characteristics were constantly displaced northward. Going downstream meant following some kind of moral slope in peoples' imaginaries as well as a sort of timeline: compared to the "wild" (*liar*) communities of the far North who were living on unregistered land that sank most rapidly, the partly regularized areas of South Kemijen were considered more advanced. Downstream neighborhoods were called "wild" if they were located on land that wasn't officially designated for residential purposes, but this categorization also drew legitimacy from allegedly more pronounced problems with flooding and contamination.

In the following pages, I explore how the marshy area of Semarang became associated with moral and social decline and how this moral geography in turn gave rise to various spatial transformations. I focus on residents with riverside dwellings or those houses within two hundred meters of the river as they were affected to varying degrees by infrastructural works on the riverbanks. When I spoke with them about the current state of Kemijen, especially the effects of tidal flooding, they often complained about the regular flooding of streets and houses, but they also remarked on how things had markedly improved. At first, this optimism, evident in the orations given at the kelurahan, deeply confused me for two reasons. First, in both national and regional newspapers, reporters and politicians were painting *rob* in increasingly dramatic tones (Lassa 2012). Second, my contacts at Semarang's Urban Planning Department (Bappeda) called the area a "problem zone." Were my interlocutors unconcerned about the increase in tidal flooding events? I knew better than to assume that people were unaware of flood risks. After all, most families residing in North Semarang had lived with floods for decades. Eventually I discovered that people were referring to specific parameters of improvement—criminality and a sense of belonging. In their accounts of the current state of Kemijen, these parameters colored peoples' notions of the past and the present. Over the years, access to medical and social services, such as health insurance, had improved; state subsidies and aid projects flowed to the area; and people had begun to feel safer. But the imaginary of the "dangerous" (*bahaya*) and "violent" (*keras*) North proved resilient. Despite improvements, North Semarang was still considered a hiding place for gangsters—as will become clear later in this chapter, the state played a significant role in emplacing this imaginary.

The floodplain is often referred to as a filthy (*kumuh*), "black" area (*daerah hitam*), a place that emanates danger. More generally, such considerations resonate with the negative stereotypes of people living in coastal regions (*daerah pesisir*). *Hitam* also refers to the putatively darker tone of their skin. In the following, I continue to query the roots and meanings of these attributes while demonstrating how river infrastructure in Kemijen, a place marked by stigmatization, plays a crucial role in various attempts to redeem the area.

In colonial times, the "chaos" piling up in the margins of the state repeatedly animated and brought questions of control center stage. As a place of informal authority (Barker 2007) and autonomy, the kampung bred apprehension among affluent urbanites and the political elite.[3] How did perceptions of the kampung become tinged with worry and trepidation? In order to get to the bottom of this stigmatization, I need to back up a little. As I showed in chapter 1, fear of excess (spread of diseases and criminals) found expression in colonial attempts at controlling the living environment of the "native" population. In 1937 a journal for Indies technicians, *Tecton,* argued that kampungs were a "source of danger. Fires and epidemics start there ('flies, mosquitoes, rats, fleas, stench, etc.') and spread beyond the kampong's limit; thieves come from there" (qtd. in Mrázek 2002, 68). Therefore, most of the Dutch interventions dating to this time aimed at opening up the kampung—by aesthetic and infrastructural means. In line with this effort to make the kampung population more visible, colonial efforts sought to remake the "dark" North Semarang into a transparent environment. Once opaque, wily waters were to be rendered manageable with hygienic care. Narrow paths were turned into wide alleys, and arbitrarily planted trees made way for curated lawns. In creating transparent and airy habitats that replaced the "dark" and putrid houses of the natives, these interventions also made the kampung inhabitants accountable to the authorities. In line with central government directives, Semarang's council "directed that the natives carry lights during the evening and night hours" (Cobban 1988, 274). The obsession with the visibility and mobility of "native" subjects meant that the Dutch intervened in indigenous residents' most private affairs. For instance, kampung inspections that took stock of native housekeeping were "undertaken in all parts of the city including the kampungs" (276).[4]

The Indonesian fight for independence and national stability required shedding the image of an impoverished proletariat and bringing the masses into unity with the nation. As a result, there was an ephemeral moment in Indonesian history when the kampung symbolized unison and hope instead of disenfranchisement and danger. Sukarno, Indonesia's first president, held that the kampung was an important anchor for the working class and its nationalistic virtues, a place from which Indonesian nationalism derived its legitimacy. Reviving the old visions of Dutch planners, who wanted to build a healthy kampung workforce—a stream of men, united by the pursuit of wealth and progress—urban planners took it upon themselves to improve the living spaces of the impoverished masses. In state propaganda, kampungs sheltered the *rakyat*—the mass of wage workers and migrant peasants in whose name the Indonesian state was proclaimed. Sukarno amalgamated the rakyat and the state in an attempt to "limit their social revolutionary impulses" and consolidate his power (Siegel 1998, 3). But when Sukarno's influence dwindled and Suharto's military regime took power in 1966, the kampung once again began to elicit fear. James Siegel delineated a discursive shift under Suharto that transformed the rakyat into a hazy mass that could hide dangerous societal elements. In fact, in the 1980s, the kampung became the stage of deadly violence against assumed "criminals." The oppressive regime behind Indonesia's New Order wanted to eliminate the subjects that once agitated against the Dutch, in order to curb their revolutionary power. While former political discourse had equated the rakyat with the nation, members of this class, along with their livelihoods and places of residence, turned into objects of public suspicion. In sum, the discourse that saw in the kampung the cause of poverty and criminality seemed alive and well in the New Order.

Throughout my research, I had numerous conversations with research participants about the Petrus (Pembunuhan Misterius, or Mysterious Killings) massacres that occurred between 1983 and 1985 and targeted "tattooed" kampung dwellers. As elsewhere in Indonesia, the typical story is that "jeep-loads of masked, armed men drove to the homes of supposed criminals in the middle of the night, abducted them, repeatedly stabbed or shot them, and left the bodies on the streets or in rivers where they became spectacles" (Siegel 1998, 2). One person told me that he was enrolled as a driver to chauffeur one such death squad through his neighborhood "once or twice." Many of my male interlocutors had memories of a frightening time when "black lists" circulated and bodies were regularly found in canals and streets. This was a time when a great many of Kemijen's current residents were adolescents or young adults. In conversations, these residents often described this period as "frightening" and "mysterious." In his analysis of the 1980s mass killings, Siegel insightfully argued that the New Order needed to control the very idea of death—a notion associated with the criminal, whose hiding ground was the kampung itself. Like these marginal spaces, the criminal was "always on the edge of Indonesian society, but never outside it" (3). Importantly, Siegel's analysis encompasses socioeconomic shifts within society that had their origins in Indonesia's economic boom in the 1970s. The New Order brought economic progress but at the cost of violent political suppression.

Memories of the killings and the criminalization of the kampung strongly inform people's attitudes toward the North. For example, Rendy, a thirty-five-year-old father of two, twice elected director of Kemijen's organization for neighborhood empowerment (BKM), considered the North morally toxic. Rendy told me that he used to be *neko-nekon* (colloquial term similar to *nakal*: naughty)—referring to his past, when he hung out with criminals, some of whom were sent to prison. He still occasionally bumped into these old associates in Semarang's bustling marketplace, Pasar Johar, where they now sold stolen goods and contraband. They always marveled at how he had made something of himself, that he had become the head of an organization. While being grateful that his situation had improved, Rendy deplored how the North had shaped his past behavior: "my environment wasn't right" (*lingkungan saya tidak benar*). Rendy claimed that the environment made him ignore, at least initially, the aid programs hosted by youth organizations, such as Karang Taruna, that tried to give young people living in poor neighborhoods a different perspective on life. A close friend and mentor eventually made him realize that his community needed him—their streets were constantly flooding. He decided to make changing the dire situation of his neighborhood his own personal charitable mission (*amal saya untuk kampung*).

For Rendy, working for the kampung embodied the "right way" (*jalan lurus*), or ethical action. As head of the BKM, he managed an annual budget of 350 million rupiah (about $26,000), and he rejected any remuneration for his work. Rendy eagerly attended most meetings of his neighborhood community (RW) as well. I remember one particular night when heads of local RTs spoke deep into the night, drinking tea and coffee on a rug laid out on the pavement between two rows of kampung houses. Rendy participated in a hot debate concerning the obligations of new residents. Should they tear and thereby nullify their old identity cards (Kartu Tanda Penduduk, KTP)—which posed the problem of dual residence and allowed for fraud—or keep them? Arief (introduced in chapter 1) suggested to withhold issuance of new KTPs if residents' old cards had not been "scratched out" (*coret*) in a correct manner. This position was challenged because this procedure would affect quite a few residents, as the neighborhood had grown considerably in the past ten years. Perhaps, one attendant suggested, they could warrant an exemption for residents whose papers were issued by the former neighborhood chief? Arief tried to abate the issue by suggesting that since the issue was governed by national (*undang-undang*) and provincial (*perda*) laws, each RT could implement these at their discretion. That stance was too ambiguous for Rendy, who demanded strict handling (*ketegasan*) of the issue. To prevent undesired people, he continued, "such as members of ISIS, from moving into our neighborhood, we have to be watchful." In Rendy's opinion, despite positive change, there was still a potential that the "dark" character of his neighborhood would remain. He admitted that residents could be "persuaded and inspired" (*bisa diajak, jadi terinspirasi*). But in his view, the transformation of the North from a dark to a socially progressive area had occurred unevenly and remained incomplete.

A conception of the kampung as a space of negative influence also came up in my conversations with Udin. Before getting married, Udin was in the habit of spending entire weekends away from the kampung in order to escape it and experience the "outside world." Trying out drugs and exploring other avenues of existence, he mingled with vagabonds in the "modern" spaces of the growing metropole, such as the central entertainment district Simpang Lima, where they would stay up all night partying. He still liked to escape from his home in the North and cruise around, a desire that is congruent with his profession as a *becak* (pedicab) driver. However, he regretted having mingled with criminals in his youth and feared that this time had taken a moral toll on him. When he was given the opportunity to become head of RT a few years ago, he refused. He confessed that he was away too often, at meetings or on the job, and felt that he was too mentally unstable to take the position. Although he was convinced that he had the skills of a good RT leader, he was afraid that he would not serve the people (rakyat) well. In fact, he didn't trust himself to be a local leader because of his temper. He told me that he often got emotional and angry (*tinggi emosi*), "from his feet to his head," when someone provoked him. For instance, when he once got into a motorbike accident, the other driver responded aggressively and started hitting him. He took a few blows, trying to calm his temper, before exploding and beating the man violently. Udin traced his aggressive behavior to his youthful engagements in the criminal world. From the moment Udin spoke of this period, he emphasized that he didn't want to become like his former friends. He simply wanted to understand how they lived and how it felt to have even less money than himself. These criminal friends of his were not from Semarang but were newcomers (*pendatang*) from Solo or East Java. Only when his friends robbed a gold seller (*tokoh emas*) and were searched by the police did he drop out of the circle. The stories of Udin and Rendy demonstrate how, and perhaps why, many in the kampung consider high moral standards to be very important for the area's positive development. Before continuing, I would like to add another story that shows the link between infrastructure and morality.

The life of Pak Marsudi, an elderly resident who grew up by the Banger River, speaks to moral transformations that resulted from infrastructural improvement. Pak Marsudi was strongly involved in communal efforts to turn the area into a livable and safe environment. The state, according to him, played a minor role in this process. He stressed that the community had built the neighborhood "on their own" and were responsible for any success trying to keep criminality at bay. By the end of the 1990s, however, the situation had become intolerable. As a result, Pak Marsudi's family decided to move to a different part of town, in fear that their oldest son would become like the many rowdies and gangsters that roamed Pengapon Street at night. In particular, they fled from the increasingly violent fights that often broke out in the vicinity of their house.

> I didn't want my children to have such a mentality. [Where we moved] there was no fighting, there were no drunks. We didn't move to evade flooding. Flooding just happens once a year. Here, if it floods once a year, that's considered natural [*wajar*]. Once a year is normal. We're flooded, yes, but actually just once a year. That's not dangerous anymore. People living here, they're in for it. We always say, we're happy anyways [*senang sama saja*].

Despite being exposed to flooding, the family decided to return to their old neighborhood a few years ago and occupy a small house off Pengapon, where they now run a laundry business. According to Pak Marsudi, who was turning blind due to an eye infection, the neighborhood had dramatically shifted, and crime rates had dropped, making the area less harmful to their children's moral development.

> It is said that criminality dropped drastically, yes. By more than 60 percent. In terms of criminals, you can say that along the river . . . perhaps 25 percent [of the criminals] remain. They disappeared, because, you know, residents were more able to control people who lived by the river. If there's someone who shows no respect [*ndobol*], [they find out] where that person could be from. Previously, people were more afraid of gangsters [*preman*]. They're not afraid any longer, because people pay attention to safety.

In Pak Marsudi's depiction, the community recognized and assumed the responsibility for handling crime and creating safer streets—not the government. However, he noted that material transformations played a central role in facilitating control: "[Before] there was no embankment, right? So, he [the criminal] tended to be down there, by the side of the road, where you can directly descend to the water. That was his hiding spot. But now, we have the inspection road [*jalan inspeksi*]. Previously, we would lose track of him down there. The sight was hindered by houses and by the public toilet building [*semacam MCK*]." Interestingly, this statement depicts the river as a magnet for criminals while also attributing infrastructural changes with returning authority to the local community.

While I was doing fieldwork in Semarang, another phenomenon sent ripples of fear over the town. Newspapers reported so-called *pegal* events on a daily basis. If I planned to visit a field interlocutor or have dinner with friends after dusk, I was regularly warned by my host family that I ran the risk of being robbed and potentially killed by merciless crowds of young gangsters who preyed on lonely drivers. Allegedly, they particularly liked to assault drivers on bigger roads that had bad street lighting. I was advised to take kampung alleyways or smaller roads, where communities had begun reinforced neighborhood watches in response to such incidents. The gangsters' acts were generally explained as desperate attempts at making a buck from selling stolen vehicles. These misled youth were seeping out from the poor neighborhoods to haunt urban spaces like major traffic corridors. Ironically, while they tended to spread out from the kampung, it was here that I would be safest. This dual imaginary of the kampung has to be kept in mind when we think about development projects. Throughout the past, kampung residents have been the subject of interventions that tried to control the dangers that emanated from and roamed this imagined and real realm. The belief that criminals continue to roam the kampung signals that preserving the positive features of the kampung and eradicating its bad ones is an ongoing process, in which infrastructure plays an important role.

Normalisasi: Sterilizing Kampung Space

The state intervention that all inhabitants of Kemijen remember is "river normalization," or *normalisasi sungai,* which was initiated in 1985. These infrastructural changes pursued under Suharto can be seen as one of the first, if not *the* first, significant infrastructural intervention into the lives of coastal residents after the colonial project of draining the swamp. The colonial intervention consisted of mainly fragmentary attempts of "opening up" the kampung and introducing Western-style drainage. While this state program purportedly aimed at reducing floods and river pollution, it coincided with the Suharto regime's attempts to bring the kampung within the purview of the state in the early 1970s. These measures were intensified in the 1980s, a "watershed period in both economic and political terms" (Barker 1998, 10) that elicited more authoritarian state measures of social control. Manneke Budiman et al. (2012, 51) argue that normalization was "part of the effort to sterilize the public space from any revolutionary potential amongst the masses." Whereas public space had been the theater of mass mobilizations under Indonesia's first president,

Sukarno, the Suharto regime instilled an image of the street as a dangerous place roamed by criminals and thugs, and discouraged people from associating themselves with it (Kusno 2000). Joshua Barker (1998, 10) sees the "criminal specter" as a "symptom of a structurally weakened state (and society) and as a convenient excuse for actions aimed at trying to overcome this weakness."

Democratization forced the state to involve residents in the infrastructural remaking of the delta. Normalized drainage infrastructure, with its various components like inspection roads, pumps, and riverbanks, can be seen as a technique of order that affords the "possibility of coordinating and increasing individual effort" (Mitchell 2002, 9). According to Timothy Mitchell, enframing further creates a division in social space and, therefore, an "ordinary way of effecting what the modern individual experiences as the really real" (60). As I will show, the method of embanking provides such vantage points, especially in the form of the inspection road, from which to observe the kampung.

Let me elucidate the concept of normalization first. A well-known example of counterrevolutionary measures is the state-sanctioned program called Normalisasi Kehidupan Kampus (NKK, Normalization of Campus Life). Implemented in 1978, the NKK sought to discourage students from "any involvement in so-called '*politik praktis*' (practical politics)" (Budiman et al. 2012). Similarly, river normalization intended to deter people from settling along rivers that were thought to constitute a realm in which subversive subjects lived and flourished. The intervention targeted so-called squatters and "illegal" residents who on top of being social outcasts due to their legal status were—and still are (see Voorst 2014, 342)—often blamed for floods. While contemporary municipal governments put forward sanitary arguments in favor of normalization, riverbank settlements are still considered a major problem. They are cast as an informal space that defies state control. Illegalization, then, mainly serves the purpose of categorizing subjects to subsequently distance them from the moral and social realm of normalized state–society relations.

As I have shown, the river normalization program had distinct historical predecessors: both colonial inspections and landscaping interventions cast northern kampungs as an "urban problem." After the defeat of the Dutch, the question of what to do with the often-degraded habitats of poor Indonesians was at the heart of Indonesia's national development agenda. These spaces were perceived as pockets of poverty and a stain on the image of the nation. Notably, the infamous Kampung Improvement Program (KIP) of the 1960s tried to tackle the problem of abject urban poverty by equipping high-density neighborhoods with new infrastructure. I mention this national program because the fact that subdistricts along Semarang's coast weren't included in the infrastructural upgrading program can tell us something about their perception by the state.[5]

With help unforthcoming from the state, residents continued to resort to informal means of making the area livable. For example, residents constructed a bridge over the Banger River. The bridge was built without substantial government help but nevertheless fulfilled the building requirements of the local government (kelurahan). I learned from Pak Marsudi that, in the 1970s, not all original buildings in the Banger delta had been residences; the riverbank was once used for animal farming. By 1980, most of the farm buildings had been converted into houses. In the following conversation, Pak Marsudi mentions the manifold purposes of the river before it was embanked.

> Initially, there was no embankment [*talut*], right? There was only a slope. So that was used by kids at the time: for playing, for gardening, [but] also as a hiding spot. From here to the south and to the north there were plants by the river. Hummingbird trees [*turi*]. They had this spiky fruit that one could eat. There was *lamtoro,* too. And they also had a function: they prevented landslides along the river. But there was even river cleaning. The [municipal] Agency for Cleanliness [Dinas Kebersihan], right, was using that long bamboo with a sieve at the tip. That was used for pulling waste to the shore. You know, before the river wasn't as dirty as today.

The river had once been a source of enjoyment and food resources. But Pak Marsudi also alludes to the riverbank as a space of concealment: here, between trees and shrubs, criminals could become invisible and evade arrest. While he is saddened by the loss of lush riverbank vegetation and a river that brought joy, he also welcomes riverbanks that are clear of undesirable subjects, such as criminals. The ambiguous nature of riverbanks, which defied the panoptic surveillance of the state (Barker 1998), was a thorn in the side of the Suharto regime. As a result, concerns of national security outweighed economic and ecological diversity.

While normalization was a top-down measure at the outset, its ideology is developmentalist. The "will to improve" (Li 2007) runs through Semarang's normalization agenda. Framed as a development project, river normalization often claims to employ modern river infrastructure and technology to improve the quality of life in riverside neighborhoods. In fact, inappropriate environmental conduct is often referred to as the primary reason for flooding before the lack of adequate infrastructure. Normalized riverbanks are thus supposed to stop pollution by residents. However, as a technocratic project focused on a dualistic relationship between kampung dwellers and the environment, normalization projects often fail to take into account other factors that increase flood risk and pollution. For example, many residents argued that increasing industrialization of the coastline was the prime cause for river pollution. As Pak Marsudi remembered:

> There was contamination from leather treatment with alkali, so that it could be used in products, like bags or shoes. Maximally, once a month they would clean [their factory] with hot water. Hot water used for flushing out the acetals would be later evacuated. . . . Compared with today, that water was quite clean. Like the river in Penjaringan [a neighborhood of Kemijen] that comes from Pertamina [an oil refining company]. That river was quite wide before, almost three meters. Its water was pristine. It would flow over here [into the Banger]. The river was very clean. Children bathed in it. It was even warm—it was wastewater from the evaporation process.

Although Pak Marsudi favorably contrasts past social benefits of the river with its present, strongly degraded state, his memories reveal an already industrially polluted environment. Residents could still tolerate these levels of pollution, but normalization tipped the scales: riverbed deepening and land subsidence made the river vulnerable to the intrusion of oceanic water. Land subsidence can be traced to ground water extraction and soil settling. Between 1971 and 1980, Semarang experienced unprecedented population growth (an average annual increase of about 45,000 individuals). Coastal neighborhoods, like Kemijen, quickly built up, accelerating soil solidification with increased ground water extraction. Taken together, ground water extracting and buildup caused a phenomenon that became known as "land sinking" (*keturunan tanah*). Normalization's focus on sterilizing riverbanks therefore screened out far more complex ecological processes.

My point is, however, that *normalisasi* had lasting and unpredictable effects on Kemijen. It changed the physiognomy of the Banger River substantially; not only was the river embanked and widened, but the government also enforced a minimum space between the bank and riverside houses for inspection purposes. This *jalan inspeksi* (inspection road) was to introduce more visibility into the densely inhabited riverside kampungs. Nowadays, the *jalan inspeksi* is one of the liveliest spaces of Kemijen. It is policed by communities who enforce traffic regulations and norms of conduct (for instance, women are discouraged from walking alone after nightfall; drinking alcohol and consumption of drugs are barred from this public space). At the same time, in many neighborhoods the *jalan inspeksi* was also appropriated not long after the riverbanks were constructed. Mostly the riverbanks were reclaimed by residents, but sometimes former landowners laid claim to a parcel with prior permission of local leaders who might expect a bribe in return. These spaces became tea shops or hang-out spots. The riverbank also features dwellings that were built by newcomers after normalization. Members of these households often don't have direct relatives in the neighborhood. Being considered newcomers (*pendatang*), they often struggle with social isolation. Building makeshift or more permanent houses on the riverbank is considered illegal but is often tolerated by the subdistrict government. While considered "illegal" residents, they still pay an informal land tax (*retribusi*). Deeply altering the kampung–water relationship, normalization extended the process of formalizing land use that began in colonial times. Both technological interventions and the extension of a state regulatory framework into the kampung were supposed to introduce a different set of human relations with the river. The "parasitic" influence of kampung dwellers on this modern infrastructure had to be curbed, by creating a proper distance between the river and residents. "Illegal" riverbank dwellers are not only discursively posited as disturbing a harmonious coexistence with water but also called out by the government for inhabiting a formally designated space.

Interestingly, normalization didn't put an end to littering. Most people, like my landlord, didn't suddenly engage in prescribed ecological practices. However, they did successfully internalize the condescending state discourse that framed their environment as filthy and undeveloped. Today, in view of continued littering, blame can be downloaded to "illegal" riverbank settlers or the government when it fails to enforce official laws. Displacing blame is evident in Eko's slur: "the people by the bridge will fish it out"—there will always be people that are even poorer and worse off than the litterer, that is, people who live directly by the water and further downstream.

While certain forms of cohabitation with the river were lost, outlawed, and criminalized in the normalization process, something new was created: tidal flooding compounded by monsoon floods. People generally hold that, in the past, flooding was tolerable. Pak Marsudi reminisced that "when the raining stopped, the flooding also stopped." Normalization streamlined drainage canals, which increased the overall water debit: "Three rivers became one: Banger River; the [East] Flood Canal and the Babon River; the one at the edge of the city. All those became one. It's like narrowing the current. Eventually, it spilled over into peoples' homes." Adin also noticed that the rerouting of Banger River led to more inundation. The river's flow used to resemble a straight line, but according to the government, it had to curve before connecting with the Java Sea. While bending rivers is in fact a well-known practice used to reduce flash floods, most residents speculate that the central Agency for Public Works (Dinas Pekerjaan Umum) acted solely in the interest of the port businesses. Bending (*pembelokan*) the Banger allowed for the reclamation of marshy land in the port area that is now used by the city's electricity plant. Notably, the swamp functioned as a protective belt from tidal waves. Adin pointed out that Kemijen's wetlands used to be shielded from the shoreline by a mangrove forest that was gradually destroyed over the course of coastal industrialization. It was a buffer zone that mediated saltwater inflow.

As Pak Marsudi remembered, *normalisasi* took a long time. Many people living on the banks of the river had to be resettled. Landowners with legitimate deeds refused to relocate. When they received monetary compensations (*ganti rugi*) from the government, they bought property nearby. The project was eventually completed well behind schedule in

1992. The remaining riverbank dwellings suggest that normalization never really achieved its ultimate goal and remains an ongoing process. Tellingly, rumors circulate that prior to normalization works, the government was too afraid of resistance from local *preman* (gangsters) to intervene. It therefore didn't carry out all planned evictions. Others believed that the government had embezzled project funds and used the *preman* story as a cover-up for running out of money. In other words, normalization neither rooted out fear and the specter of criminality nor achieved public monopoly over urban space, which is why the government continues to push normalization plans. What normalization did manage to achieve with success was to reinscribe the dominant image of deviation and backwardness upon the North. I now turn to a government-sanctioned river imaginary that showcases attempts by residents to change these images and moderate state fears around banjir. Here, the enrollment of the normalized riverscape in festivals is a way of linking locality to the state's modernization plot.

Normalisasi and Local Festivals

Since 2000, areas affected by strong tidal flooding have garnered increasing attention of both the media and local government—which, as I learned from local journalists and residents, often entertain strong nepotistic ties in Semarang. In an attempt to focus the energies of municipal politicians on deficient river infrastructure, residents of Kemijen have repeatedly organized events that explicitly revolve around the river and flooding. In 1998 a river-cleansing ritual (*ruwatan*) was organized by inhabitants of Kemijen. This event also marked the launch of the resident organization Komayu, whose subsequent and ongoing river-focused activities I describe in chapter 4. In 1998 Indonesia's longtime president, Suharto, was forced to step down in the wake of mass protests against his corrupt and self-serving administration. Yet, as this section illustrates, residents continue to stage river festivals in an effort to renew governmental commitments to normalization, a formerly authoritarian urban project launched by the Suharto regime.

The cofounder of Komayu, Adin, explained that *ruwatan* (cleansing) ceremonies originally served to "sidestep dangers or disturbances" (*meminggirkan bahaya atau gangguan*). In 2005 and 2006, they organized *ruwatan* along the Banger River. It was thought that these cleansing ceremonies would help prevent flooding. While a religious figure conducts such ceremonies, they also involve the ordinary population. The newspaper *Suara Merdeka* stated that families placed bamboo sticks (*obor dari bambu*) on the riverbank. While having spiritual overtones, these ceremonies were vehicles of political expression; they served primarily to "push the government" (*mendorong pemerintah*), as Adin said. He therefore also called the event a "demo" cloaked in culture. This "demo" served to remind the "people's delegates" (*wakil rakyat*) of society's problems (*permasalahan masyarakat*). A news article actually put the term *diruwat* in quotation marks, probably to underline the slightly quirky use of ritualistic language for clearly political purposes.⁶

The 2005 *ruwatan* failed to attract substantial attention from Semarang's government. The workshop that residents organized to discuss the continuation of normalization was poorly attended and apparently of little appeal to politicians and bureaucrats. The festival, however, continues to play a role in civil organizing to this day. In 2006 and 2007, the local NGO Perdikan co-organized "climate change festivals" with riverside dwellers using funds from the international NGO Mercy Corps to raise awareness around the potential impacts of climate change. Again, the event had infrastructural connotations, as these events, sometimes also referred to as "festival of the people" (*pesta rakyat*), specifically drew attention to flooding. Furthermore, in 2008, engaged residents organized the first "boat competition" (*lomba perahu*) on the Banger River. In terms of outcomes, festivals always led to some form of improvement, as Adin noted. For example, embankments were often repaired after a festival.

In sum, these festivals aimed to remind the government of unfinished infrastructural improvement works begun in the past. More specifically, in the context of cultural festivals, residents would ask the government for the construction of floodgates and river walls along the whole river. Festivals often occurred and continue to take place on the *jalan inspeksi,* which demonstrates a local embrace of normalization. Thomas Blom Hansen (2001, 53) has written about the rise of similar festivals in postcolonial Bombay. In Bombay, such public festivals "are closely tied to the locality's wider politics of representation, [and] to the production of the neighborhood." To produce a neighborhood is, according to Blom Hansen, "to claim a certain identity, a belonging, and thus, by implication, a set of entitlements for a particular area and for the people living there" (54).

In 2015 news broke that the city government would expedite a delayed antiflooding project (the polder project I analyze in the final chapter) by using municipal funds. The project promised significant improvements of the Banger's infrastructure. In view of this impending feat, the mayor was going to visit Kemijen—a highly important and anticipated event, as the opening vignette of this chapter noted. His visit was supposed to coincide with the celebratory announcement of plans for a kampung beautification project. For this reason, residents cooperated with the kelurahan in order to organize a festival, the Caring for Banger River Festival (Festival Kepedulian Kali Banger). Tellingly, the organizers explained to me that in previous iterations of the festival, they simply called it Festival Kali Banger. But that title only related to the betterment of river infrastructure. This time, they wanted to show that humanity could also be "built" (*tapi manusia juga dibangun*). The "caring" in the title highlighted the human role in infrastructural maintenance, a tenet of local worth since normalization. Building on spaces and infrastructures that were created in the wake of normalization, these festivals cast normalization as a desired, participatory, and resident-driven project.

At another neighborhood meeting aimed at preparing RTs for the mayoral visit, the lurah held forth a zealous appeal that underlined the role that residents' behavior (*perilaku*) played in improving the neighborhood. Although the community had shown exemplary commitment and willingness to cooperate with the local government, residents had to be sure to send the right message on the day of the mayor's visit.

> Concerning the mayor's visit for the river cleaning, our collective behavior has to reflect a clean Banger River and local youth. . . . I want to show that the residents of this area care about waste— that's where I'll be pointing. Sure, our drainage system is still not working . . . but garbage makes *rob* look even worse. . . . Those are old habits but they have a strong impact. I want residents to look united [*guyung*] and motivated so that we come across as

residents that are actually interested in creating a better environment. The mayor told me that if he gets reelected, we will receive more help.

The lurah asked residents to refrain from complaining about the current situation and the still deficient drainage infrastructure. Instead, he encouraged them to show commitment to the city's drainage plans, as this, he assured, could set processes in motion that might cure the system's dysfunctions. "Caring," as opposed to indulging in "old habits," is the effect that festivals and other public events seek to showcase. The speech and festivals urge residents to distance themselves from the past and those that still inhabit it, such as "illegal" riverbank dwellers. Embracing the drainage system is a prerequisite of local progress and a process that is mediated by caring for infrastructure. Normalization has therefore seeped into social practices and peoples' outlooks on the future. Though infrastructure isn't perceived as a panacea, improvement is contingent on building a good moral character and showing care for the neighborhood's environment.

Epilogue: "Killing the River"

A 2012 *Jakarta Post* article paints the future of Semarang's low-lying delta in apocalyptic tones: "Like Venice and Mexico City, the city of Semarang . . . is slowly sinking. Rising sea levels as a result of global warming further add to the burden of these cities" (Lassa 2012). Expected sea level rise casts a dark shadow upon the future of Semarang's littoral. According to vulnerability assessments conducted in northern kampungs and frequently cited in journalistic reports, residents "experienced" increasingly high rates of land subsidence, "ranging from about 10 to 15 centimetres annually" (Lassa 2012). In his article, reporter Jonatan Lassa tries to capture reactions to the nearing catastrophe:

> For Mrs. Elizabeth, who was born [here] and lived in the village for 43 years, the problem of the sinking houses is caused by the "rising water" coming from the sea. However, she didn't make any connections to global warming as some people in the city might have. We asked a group of boys in the village what they thought caused houses to sink and their answers were similar to

Mrs. Elizabeth's. We further asked the boys: "Which one is true: Is it the sea water that is rising or is it the houses that are sinking?" The boys apparently became confused and shrugged their shoulders.

The newspaper article seems to enlist these kampung inhabitants as extras in a familiar media staging of Semarang's backward North. The stereotypical country and city divide is upheld as residents come across as ignorant and uninformed, whereas "some people in the city" have already made the connection. The article goes on to suggest that the government is not prepared to wait for residents to grasp the gravity of the situation. Based on scientific "evidence," such as Intergovernmental Panel on Climate Change (IPCC) reports and vulnerability assessments conducted by foreign experts, Semarang's government is indeed poised to undertake a range of infrastructural projects that aim at protecting the littoral from floods. The plans foresee the fortification of the coastline by installing breakwaters off the coast and building seawalls along the shore. Another step in this process is the closure of river outlets to stave off river–ocean interaction. The municipality pays external experts to monitor river debits and simulate flooding forecasts. One of its trusted advisors, Pak Suseno, strongly suggested the implementation of multiple polders (closed hydrological systems) in order to fully regulate water flux throughout the city. He liked to call the act of poldering "killing the river" (*matikan sungai*), which underlines the kind of radical intervention into the urban ecology that he had in mind. For Pak Suseno, killing the river meant engineering water flow and gaining full hydraulic control over water. The figure of speech was not an exaggeration: turning all of Semarang's rivers into polders would literally be a death sentence for a long-standing principle and foundation of urban life in Semarang— polders would radically change the relations that people entertain with nonhuman entities, such as rivers. A new urban life would be built on this murder.

The perceived *rob* crisis has renewed commitment to shaping the North by infrastructural means. The scope of these suggested interventions is nothing less than monumental. Incrementally, the shore will be fortified and rivers running through residential areas will be dammed and re-embanked. Infrastructure changes will coincide with crucial adjustments in the "social" and "cultural" spheres of floodplain communities: engineering flood control requires educating riverside dwellers so that they treat rivers as delicate parts of an organic system.

In 2014 Bappeda crafted a new drainage law that can be seen as a roadmap for building a coherent and integral drainage system. The policy envisions new roles for civil society in maintaining and monitoring drainage infrastructure. Section 6, on *pengelolaan* (management), outlines the governing role of a so-called local management body (*badan pengelola*). This civil institution is supposed to assist government agencies in operating and supervising the system. The role that residents are meant to play is further specified in paragraph 59: "affected," that is, flood-prone communities, are given the "opportunity" to join in the supervision of the drainage system by forming "working groups" (*kelompok kerja*). These groups, consisting of flood victims, are expected to funnel communal aspirations and "formulate ideas" (*merumuskan pemikiran*) for the government. In a conversation with a lower-rank member of Bappeda, I learned that this section was arguably the most innovative addition to the city's set of drainage regulations. However, the agency had admitted to difficulties in imagining the "way" in which the "working groups" would cooperate with the government. It was therefore easier to imagine the demise of Semarang's riparian system than assign society a role in its future governance. As the employee added, the precise kind of public involvement would have to be specified at a later stage through another bylaw (*perwal*).

Publicly, the new drainage law was heralded as a symbol of rare prestige. Semarang was considered the first Indonesian city to adopt such "participatory" measures. While the plan has not yet been realized, imagining a future has already translated into concrete changes in the urban fabric. Bappeda recently helped bring a much-awaited polder project to fruition (the spatiotemporal and political effects of which I describe in detail in the final two chapters). Suffice it to say that this pilot polder project was supposed to pioneer a novel form of association between government and society not just through but in concert with infrastructural modernization. Partly improvising on the new regulatory environment created by Bappeda, the polder project managed to assemble diverse actors—residents, city workers, and Dutch engineers—who devised ways to put residents in charge of drainage infrastructure.

It was no coincidence that the consultant hired by Bappeda to review the drainage plan was also chairman of the advisory board created to guide the implementation of the polder. The new drainage law was in part pretailored to the legal requirements of the polder project, as it would require raising a "water tax" from residents. As the chairman pointed out to private stakeholders at an informational meeting: "when the drainage law is out later on, we'll have the right to take contributions."

In a private interview, the polder board's chairman said that Polder Banger is the beginning of a complete refashioning of water management. All rivers will be fully embanked all the way from their upstream source to the littoral (Figure 3). He likened these new embanked and fully controlled rivers to *jalan tol*, or toll roads, as they would transport surplus water to the estuarine. His comparison could not have been more telling. Similarly, Indonesian scholar Abidin Kusno (2000) has analyzed the New Order imaginary of urban flyovers—throughways that allow for unobstructed flow of cars within cities. He explained that modern

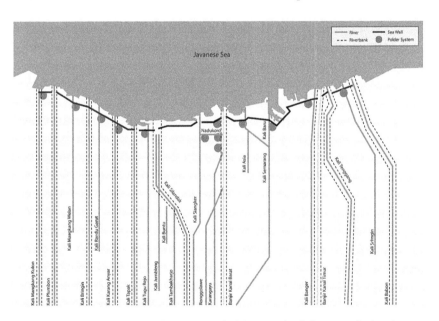

Figure 3. Drainage roadmap for Semarang's delta. Nearly all rivers are depicted as framed by riverbanks (*tanggul sungai*) and part of an integral water management system. The plan envisions turning eleven out of twenty-four urban waterways into polders. Map by Sol Iñigo.

urban infrastructure required residents to develop an infrastructural aptitude, thereby constituting specific political subjectivities. What congeals here is an expansion of state authority into the everyday lives of kampung dwellers through participatory water management. However, as I will show, the vehicle that would have allowed this rearticulation of state, residents, and water infrastructure—the Banger Polder—didn't materialize as planned.

The drainage plans of the government demonstrate that dreams of taming and channeling water are still being dreamed. While the political conditions of their articulation have shifted, they stem from concerns that are often still the same as in colonial times. Enrolling the community in new programs that abide by "normalization" principles follows well-rehearsed arguments that continue to patronize kampung inhabitants and see in their environment the reason for moral and environmental misconduct. "Killing the river" (*matikan sungai*) indeed means carrying out a specific incision into the body of the kampung, a problematic extremity. As this chapter and the previous one demonstrated, the environment and the kampung (subject) have been deeply entangled in state discourse since the beginning of Semarang's urbanization. In the colonial era, Semarang's city council aimed at modernizing the "dark" North Semarang and in so doing laid the ideological groundwork for subsequent technological modernization ("normalization") schemes aimed at reshaping residential conduct. Today, it is through the "democratization" of water management that the parasitic relationship between coastal kampungs and the city's drainage system is supposed to come undone. By pointing to the environmental impacts of normalization and its ecological consequences—tidal flooding—I have outlined the flipside of this dream. The next chapter will look specifically at present struggles to deal with the outcomes of normalization.

The delta of Semarang is still in the making. The findings discussed in this chapter suggest not only that the kampung and the river are contested objects, subject to shifting economic and political interests, but that both the government and the population perceive them as unfinished. Conjuring up commitment to development and environmental "care" has become a central interface through which inhabitants and the government set the terms of these ongoing transformations. I have outlined some possible reasons for this sense of incompletion. First, residents have turned paternalism and state anxieties into catalysts driving the infrastructural improvement of the area, in which some residents, such as Rendy, find an active role. Second, the government is striving to contain potentialities: trickles of immigrants, crime, and water that could turn into a flood of sorts. As a result, kampung inhabitants' marginal position with regard to the political processes and powers that shape their environment, its infrastructures and biophysical processes, rematerializes in state programs and infrastructural schemes.

3

FLOATING
ENDURANCE AND THE "QUASI-EVENTS" OF LIVING WITH FLOODING

"We are stranded in the middle of the village while the water level keeps rising." So begins the Dutch-produced film I watched at a meeting between Indonesian provincial agencies and Dutch NGOs and consultants.[1] The documentary-style film argues that the destruction of mangroves and coastal habitat is the root cause of increasingly frequent tidal flooding in Semarang's neighboring town, Demak. According to the Dutch organization EcoShape, the group behind both the film and a pilot project aimed at rehabilitating Demak's mangrove forest, this ecological crisis requires urgent intervention.[2] Many of the meeting's attendants had also been involved in workshopping solutions to Semarang's tidal flooding issue, which, compared to Demak's problems, seemed much more complex. Demak's coastal environment appeared curable. Existing environmental damage seemed reversible, and future risks could be averted. Such "natural" solutions were not available in Kemijen. Here, mangrove restoration had become unimaginable, as the space once occupied by mangroves is now home to thousands of residents, businesses, and communities. What remained possible seemed like only a set of unsatisfying and very difficult interventions. At the end of the meeting, I spoke with a member of EcoShape, which was also involved in rehabilitating a sizeable stretch of coastal mangrove forest just outside Semarang. Intrigued by the surging presence of Dutch development projects in Semarang (see also Yapp 2020), I asked her why her team had decided to work on the margins of Greater Semarang. Her answer was straightforward: "The coast of Semarang City is too *hard* for this kind of project" (my emphasis).

The poem "Semarang Surga Yang Hilang" (Semarang, a lost paradise), by Djawahir Muhammad, also speaks to physical mutations that make North Semarang *a hard* place. In addition to describing the metamorphosis of the northern swampland into urban land, he offers a poetic description of how these transformations have made and continue to make life on the coast *hard*: "Sticky air, dirty skies, stuffed gutters // traffic jam // Nature has become savage." As such, the poem initiates this chapter's reflection on the hardship of dealing with an everlasting present textured by chronic disaster. Throughout my fieldwork, I began to understand hardening as an ongoing, layering process that forecloses futures in specific ways and forces individuals to deal with the present through disparate strategies and distinct architectures of time (Sharma 2014). In this chapter, I describe the experience of a temporal and existential cul-de-sac composed of chronic infrastructural deficiency and breakdown. I further illustrate how a complex array of infrastructures, largely remnants of previous plans gone awry, limit people's agency to create long-term plans for a better future. In the locally famous poem, Muhammad describes the dawn of a coastal settlement in which people lived harmoniously with the delta's nature: a nature that was benevolent, malleable, and pregnant with possibility. Throughout the poem, the passage of time flows harmoniously with change and ecological metamorphosis, until suddenly the nexus between time and nature stops working for its human inhabitants. Having been enclosed and commodified in the wake of industrial capitalism, nature grows "savage" and rebels against humanity. Muhammad, who grew up in North Semarang and calls himself a witness (*saksi*) of the area inhabitants' plight, depicts the lives of poor residents as stagnating in a toxic mixture of air pollution and filthy flood water. His words convey a sense of being stuck in a stretched-out present, one that feels like an extended apocalypse yet lacks the usual concluding catharsis. After describing the effects of normalization in the previous chapter, I will now focus on its aftermath, including the stench of stagnant outflow and the repetitiveness of development projects that fail to lead to improvement. In Muhammad's poem, people's lives end up enclosed in an era (*zaman*) wherein an exact ending seems uncertain.

In the following, I show that the process of river normalization has resulted in the proliferation of individual plans to reduce the hardship of flooding. Richard Baxstrom (2012) has coined the concept of the "baroque mode" in to order to demonstrate an urbanism that forces subjects to strategically orient themselves to plans and maps. This form of urban living doesn't allow subjects to come to rest and forces them to constantly recalibrate plans and practices to shifting urban futures. Similar to the interlocking development schemes that, according to Baxstrom, shaped the urban fabric of Kuala Lumpur in the 1990s, river normalization introduced new realities and temporalities into the lives of northern dwellers. But above all, it produced contingency and socio-material incongruities. The baroque mode of living manifests in an idiosyncratic set of plans that attempt to govern the future by attending to the multiple "twists and folds of the present" (137). This orientation to plans allows residents to fold ruptures, variably experienced or witnessed due to evictions, house erasures, or floods, within their actions and movements. While attending to the twists and folds of the baroque mode, I'm interested in examining the subjective experience of urban worlds that Filip De Boeck (2014, 80) calls "in crisis." De Boeck argues that people in Kinshasa and elsewhere "have no choice but to continue to live in a world that seems to be falling apart before their very eyes" (80). De Boeck queries if and how people come to terms with the many, "daily experienced ruptures and breaches in their lives" (81). In North Semarang, it is often government plans that cause such ruptures. One way in which ruptures get folded into people's lives is through repair. Here, I rely on Steven Jackson's (2014) critical notion of repair as that which maintains the extant yet failing systems of modernity. In view of constantly required repair of domestic and public kampung infrastructure, I suggest we understand both the "baroque" disposition and the repair as integral expressions of the chronic.

Decomposition and Repair in the Meantime

De Boeck traces the popularity of apocalyptic visions of time to a quest for completion; in contrast to the chaos of everyday life in the city, the apocalypse is envisioned and longed for as the end of breakdown and a path to liberation. Notably, the decomposition of people's lifeworlds in Kinshasa coincides with a war-inflicted memory crisis—a breakdown of the production of history—which results in the "impossibility to place or *posit* death" (De Boeck 2014, 81). The present moment, in which the idea of doom is a function of hope, turns into a stretched interlude, a meantime, with death embedded in its diffuse time scale: "not only is death no longer as before, it is no longer the end of the world either" (De Boeck 2005, 25). In this extended moment of breach, the absence of reliable historical narratives ultimately renders impossible the proclamation of an era's death. In the previous chapter, I showed how river normalization, a colonial vehicle of modernization, never actually ended. While being packaged in a new developmental order laid on top of colonial hierarchies, it does not realize the promise of modernity. In the wake of failed national development, De Boeck (2014, 108–9) observed that people try "to escape the contradictions and the disjunctures" of this postcolonial experience by reviving communitarian ideals. This exit strategy, however, is systematically frustrated by the contradictions of postcolonial reality. For the purpose of this chapter, I focus on attempts to endure the disjunctures of the postcolonial experience by pointing to the revival of pumping communities (*pompanisasi*) and the hopes articulated with the realization of a municipal polder project. However, I relate these attempts not to the promise of apocalyptic deliverance but variably to a religiously motivated work ethic and the promise of economic progress.

According to Jackson (2014), (infrastructural) repair invariably constitutes an aftermath, an activity following the failure of technological systems that are embedded in social worlds. Susana Narotzky and Niko Besnier (2014, S8) have argued that, when dealing with a world in crisis, actors draw on the resources of an environment that is "largely not of their own making but in which they have to live." Repair draws together limited resources, such as time and material. In a similar vein, De Boeck (2014) considers crisis as a catch-22 situation, which affords possibilities for action, yet none of them are satisfactory. I rely on De Boeck's catch-22 conception of crisis to describe infrastructural breakdown in northern kampungs. Northern kampungs attract a sizeable amount of small-scale infrastructure projects and a considerable influx of materials. Gravel, cement, brick, sand, and labor keep flowing into the kampung, but instead of making inroads in reducing flood risk, largely uncoordinated projects fizzle out or, even worse, tighten the grip of toxicity and disease.

The premise of this chapter is that people don't cease to exist or act in the face of this dilemma. On the contrary, there is lively movement within people's experience of an everlasting present. This steady movement produces a differential buoyancy, an equitably distributed ability to "float" or stay afloat. From within the crumbling postmodern swamp, projects can emerge that suspend decline while also confirming this trend. Povinelli qualifies these projects as quasi-events because they are mostly misrecognized and overlooked. Quasi-events can be the result of an intentional consolidation of energies (labor, capital, or material resources), but they are prevented from passing critical thresholds of "eventfulness." The critical evaluation of what counts as eventful has to be understood in relation to the temporality of late liberalism, according to Povinelli. Action, here, is rhythmed to specific regimes of social belonging, such that "people experience the kinds of events that make up their lives not only as ordinary but generalizable" (Povinelli 2011, 133). Quasi-events are costly and have consequences. In particular, Povinelli pays close attention to what she calls the violence of enervation, a process that leads to bottomless exhaustion. In the realm of infrastructural projects, materials and good ideas always seem to accumulate but never in a productive fashion. When such projects eventually get cut off at the pass, they will have used up all reserves of attention and energies from exhaustible bodies, often along with all obtainable material and financial resources. The various ways in which forms of eventfulness distribute the feeling of enervation and endurance demonstrate why certain individuals' suffering goes unnoticed or is taken for granted.

In the following section, I describe the experience of staying afloat from the perspectives of various local residents. For these people, recurrent events of tidal flooding demand staying tuned into environmental and infrastructural repair projects and local weather conditions in order to successfully improvise timely adaptations. This also means mobilizing energy for their own projects of repair and flood prevention. However, tidal flooding does not form a window into the long-term future of the neighborhood, its relations with the wider city, and the place occupied by the people living within it.

In the photo-ethnography part of this chapter, I examine specific techniques of recursive flood prevention at the level of the house(hold). These techniques—"architectures"—demonstrate the residents' relations with the temporalities of development plans and their outcomes, and the uncertainty these introduce into decision-making processes. Drawing upon firsthand experiences of residents and observations of the rhythms of everyday life, I capture the erratic, "savage nature" of North Semarang. I pay particular attention to two northern neighborhoods that belong to different yet neighboring subdistricts divided by a main road. While these neighborhoods have developed in the "margins of the state" (Das and Poole, 2004), this is not to say that they are autonomous. Rather, I hold that kampungs have been assiduously constituted by different regimes of power, both colonial and postcolonial, that capitalized on unequal development. Today, the Indonesian state efficiently relies on kampungs as partly self-governed housing projects for the workforce—auto-constructed social housing, if you will. By withdrawing from the scene of the kampung, the Indonesian state makes its otherwise ubiquitous presence less felt (Newberry 2008). While this chapter doesn't speak directly to this kind of state presence or the role of state programs in giving shape to local projects, the following chapter homes in on the claim-making strategies of particular residents in order to illustrate state–society relations.

Act I: Bric-a-Brac

On many occasions, I sat with Deni and his family in front of their riverside house, breathing in the smelly exudations of the Banger. On rainy days, the humid kampung air can reek of sewage and wet debris flushed into the Banger River through polluted drainage capillaries. Deni is married and has three children. His youngest son, Putra, is a lively, constantly babbling boy and superhero fanatic. He is a source of entertainment for the whole family and neighbors, and is rumored to have the gift of seeing ghosts. The family's small brick house, inherited from Deni's parents, is separated from the river by a narrow embankment road mainly used by pedestrians and motorcyclists. When I visited Deni in the evening, we would chat at his food stand by the main street. Here, heavy trucks of the state-owned oil company Pertamina and travel buses would roar past us, while thirsty mosquitoes whirred around our heads, attracted by our sweat, the smell of sugary *teh manis* (sweetened black tea), and the blue hue of the energy-saving bulb hanging from a cable. Most of Deni's routines take place between the river and his food stand, which are connected by a short, paved alleyway (*gang*). He often contrasted the smallness of his everyday life to the excitement of his past life as an English teacher and tourist guide. Now, he often said, "I can barely remember English but it's like that." On occasion he would ask me for the phone numbers of other Westerners so he could offer them tours to Java's cultural sights.

Deni's oldest son, Toni, who has ambitions to study abroad, once told me that wealthy neighbors have already moved uptown (*naik ke atas*) to escape the floodplain and socially stigmatized area. Economic success in Semarang often translates into social *and* geographical ascension. If residents have the means, they might decide to leave their downstream residences behind. Toni deplored that the few successful residents stopped caring about the floodplain kampungs after they departed, as if they wanted to excise this place from their identity and present. Deni himself regularly complained about lacking public commitment to improve water infrastructure in his subdistrict, and a dearth of helpful public figures (*tokoh masyarakat*). Getting worked up, he lamented that nobody stepped up for them, not even the wealthy neighborhood chief (RT), in whose flood-safe house I was renting a room.

Deni's family was particularly worried about high tides during the dry season, when the embankment (*tanggul*) barely withstood the additional pressure of in-flowing tidewater. At high tide, the Banger looks like it is on the verge of bursting. Because the river water doesn't visibly flow, the river's physical mutation is better approximated by the notion of gradual swelling. According to Deni, residents have informed the subdistrict government of the riverbank's damaged state on numerous occasions, but nothing has been done. The dirt floors of his house are flooded almost daily due to the river's tidal expansion. His family resorts to stacking miscellaneous building material, like bricks, against the bank in hopes of increasing its stability, but the water still somehow finds its way through cracks in the wall and pavement. With flower pots placed on top and red bricks stacked against it, the riverbank looks like a shelf cluttered with bric-a-brac. There is something romantic and nostalgic about the space between the house and the riverbank. The low frangipani plant in front of the house provides pleasant shade. Deni's daughter, Eny, seemed to especially enjoy this space in the morning, when the air was fresh and breezy, providing an escape from the moldy inside of the house.

Not only in Semarang, but also in other Javanese cities struggling with increasing tidal highs and land subsidence, tidal flooding events are called *rob*. Upon visiting the street in front of Deni's house, a resident of the flood-prone area once told me that most residents welcomed *rob* for two reasons. First, its regularity allows for reasonable predictions of flood risk, as people know what normal and abnormal river levels amount to. For example, if *rob* is strong, coincidental rainfall will cause rivers or canals to overflow. Second, tidal floods provide *rezeki,* a term that can variably mean livelihood or luck. When I asked him what he meant, he pointed to the streets. Lifting them required both labor and local oversight. *Rob* is the pulsating vein of economic and social life in most of North Semarang. In the presence of flood risks, residents have made their lives, pursuing their own projects with more or less success. More importantly, however, floods are not regarded as necessarily impeding economic or personal success. With the regularity of *rob,* success in an economically stagnant area, with outdated infrastructure and a rapidly shrinking built environment, depends on harnessing new entrepreneurial strategies. Welcoming *rob* is therefore not the same as normalizing crisis and risk. Instead, it is a stoic stance through which one attempts their best to succeed or wait for better times. It is this generalized distribution of risks and opportunities that frames efforts to prevent and live with flooding as a quasi-event—something that puts a repeated strain on the lives of dwellers without essentially disrupting it. Povinelli (2011, 4) frames the suffering of marginalized subjects in late liberal economies as a series of quasi-events through which their lives digress into a "form of death that can be certified as due to the vagary of 'natural causes.'" This kind of "dispersed suffering," she argues, depends on socially consubstantial ethical responses to such suffering. For example, when Deni complained one day about chronic bleeding from his anus, Eny joked that he always liked to whine. When Deni added that he also regularly had fevers, his daughter laughed. Undoubtedly, Deni's physical suffering was compounded by frustrated attempts at protecting his house from floods. His actions were determined in relation to the temporal structures of neighborhood and large-scale plans, such as infrastructural improvement projects. His daughter mocked his ailments as petty and unwarranted, as if they didn't deserve the same attention as protecting their house and safeguarding his children's future. Baxstrom (2012) considers plans as events that respond to other, more prominent events, such as institutionally driven or dominant development schemes. He shows the abilities of individuals to sense and respond to the multiplicities and complexities of daily life in urban places (around the world). When individuals' plans fail, though, how do we account for the fallout?

While Povinelli holds that dominant plans can provide the stuff from which the scaffolding of individual "social projects" are made, she also shows how such projects are rarely voluntary. Deni's family's attempt to stop flooding by stacking found bricks against the river wall disaggregates aspects of the material world. This world itself is a result of dominant plans, which materialize unevenly and treacherously in their lives. They provide, however, fragments from which social projects that produce momentary relief can be mounted. These new material scaffoldings of "endurance" are, however, exposed to multiple stresses and often don't last long. Instead of "making anything like a definitive event occur in the world" (Povinelli 2011, 10), such attempts quickly short-circuit; they can therefore be understood as efforts, in an individual and collective sense, to endure ongoing structural harm. Despite failing to disrupt social grids of time, these efforts exist in concrete ways that express modest aspirations. While Baxstrom (2012) captures innovative and congenial responses to development plans, he doesn't account for their demoralizing effects. Quasi-events rely on an ethics that downplays the suffering that results from failed projects, as expressed in Eny's critique of her father.

When I walked by Deni's home on a random afternoon, I saw the whole family sitting by the river. As I stopped to say hello, I noticed that workers (*tukang*) were busy dredging the gutter (*saluran*) in front of their house. Wider and deeper, the gutter now looked like a small water conduit with defined edges (Figure 4). Putra was absorbed in overseeing the works and ran around cheerfully in his Spiderman pajamas. Deni explained to me that they were carrying out kampung improvement work (*kerja bakti*) as they had received IDR 16 million (ca. USD 1,200) from

Figure 4. Canal digging in front of Deni's house. Photograph by the author, 2015.

the government for repairs. He added proudly that he had been elected substitute neighborhood head (*wakil RW*). Repairs were deemed necessary because the gutters were almost completely clogged. After the deepening was complete, water and waste could run more easily to the station where it was pumped into the river, Deni explained in a workmanlike manner. We agreed, however, that they still had to raise their floors if they wanted to be out of *rob*'s reach. In fact, that was why they saved the brown soil from dredging the gutter in an adjacent roofed part of their house. Eny said smilingly that this free dirt (*tanah uruk*) will come in handy in the future.

In view of the importance of repair in a world based on creative destruction, Jackson (2014) has urged scholars of innovation and media to rethink the relationship of technology and infrastructure to broader social worlds. He challenges us to take erosion, breakdown, and decay as a basis in thinking through technology. "Broken world thinking" questions the belief that only innovation and construction provide the way forward. Infrastructure, the scaffolding of progress, has taken a crack. This is nowhere more visible than in Deni's case. Cracks show in the embankment and road in front of his house, and seem to stretch back as far as the facade of his own broken-down abode. Jackson suggests paying attention to the manifold and ongoing activities by which some kind of stability is maintained in this world of breakdown. These invisible arts of repair restore infrastructure, "one not-so-metaphoric brick at a time" (Jackson 2014, 222). Rearticulating materials to achieve stability can occasion hope, as Deni's daughter's smile suggests. In this world of suffering and possibility, repair is a labor of care "by which order and meaning in complex sociotechnical systems are maintained and transformed" (222). Jackson also points out that rearticulation falls under "practice" and not "representation." We can observe this in how Deni's activities of repair are deemed undeserving of representation. They count as cunning and unofficial; while some residents shrugged them off as inefficient, officials I spoke to suggested that they might even undermine the original principle of river infrastructure by disturbing its technological logic. But in a world that is always breaking, generating excess and risk, Deni and his family have to put up with such accusations. As Jackson puts it, "repair inherits an old and layered world, making history but not in the circumstances of its choosing" (223).

Act II: The Neighborhood Pump

Another daily interlocutor of mine was Arief. Although not native to the area of Semarang, Arief devoted much of his leisure time to maintaining order and cleanliness in his adopted neighborhood and home. "This is my turf [*wewenang*]," he once said laughingly, gesturing to a stretch of about one hundred meters along the embankment. Even before being elected as a neighborhood head (*ketua RW*) three years ago, Arief felt deeply responsible for the cleanliness of his "turf": he swept the street separating the river and his house daily, he cleared waste off the gutters manually, and he assigned space for garbage disposal. He attended most meetings at the neighborhood level and convened the RT heads every month. Arief's commitment to cleanliness, civil engagement, and honesty has produced tangible results in his eyes. He was convinced that his turf was the cleanest in the whole subdistrict. In fact, when the mayor announced an official visit shortly before Ramadan, the subdistrict head (lurah) suggested a solemn inspection of Arief's territory, which served as an example of successful local governance. Volunteer work often left Arief exhausted to the point of passing out during conversations. When we attended neighborhood meetings together, Arief was often the first to doze off on a plastic chair in the second row. His wife, Ariel, was sometimes worried about his permanent somnolence. During the day, Ariel sold snacks and beverages in a makeshift canteen by the river while Arief worked in a local baseball factory. His family relied on his daytime job earnings, which amounted to minimum wage.[3] After returning from work in the early evening, he would check in with his family and closest neighbors, have chats, and then help out in the canteen. His devotion to his community overlapped often with his snack-selling business: many older constituents of the neighborhood association (RW) would come by to chat, hang out, and eat, which is why their canteen was quickly nicknamed *dewan* (parliament or council) after it opened during Indonesia's presidential elections in 2014.

Located on the riverbank and about one meter below street level, Arief's house is flooded every other day (Figure 5). Then, the floor tiles were inundated—chairs, benches, and dressers immersed in a brownish liquid. According to Arief, it used to be even worse. Two years ago, Arief's neighborhood therefore purchased a pump with a pro-poor government

Figure 5. *Rob* in Arief's house. The water remained trapped for about one day. Photograph by the author, 2015.

grant. If it was midday, that diesel pump would usually be running to bring down the water level in the street gutter. The pump dumps sewage water and runoff into the river. During high tide, however, *rob* prevented them from pumping surplus water into the river because it could not absorb it. "It makes no sense to pump out the water now as it would return just as quickly," Arief explained to me, indicating the uncertain usefulness of their collective effort. This year, his family was able to save enough money to raise the floors of their bedroom so that they would not have to sleep in a dark puddle. But the other rooms would have to wait. Private investments and communal efforts provide relief from floods, but it is rare that the circumstances allow them to be coordinated.

A few years ago, Arief started advocating for the aforementioned communal pump. His engagement as neighborhood chief earned them the cherished diesel pump, which they run as often as possible. Arief suggested a nonbinding monthly community tax (*iuran*) to pay for maintenance and operation expenses—a rule most of his constituents gladly comply with. For now, *rob* was "under control," many claimed, since they were taking water management into their own hands. When I asked him why he invested so much personal time and labor in the neighborhood organization, he answered:

> I tell my friends not to expect help [from the government]. Poor people [*orang tidak mampu*] ask for help. Help comes only once. But we work in the name of God [*ibadah*]. Our thinking has to be focused on the long term. In the past, we didn't have our own pumps [*pompanisasi*] and everybody had to see for themselves [*masing-masing harus bertahan*].

In view of increasingly frequent occurrences of *rob* flooding, Arief distinguishes philanthropic support offered by state programs from long-term investments done in a communitarian fashion. He also contrasts sporadic repair (of neighborhood infrastructure), the kind of work on which Deni's existence depends, with schemes that put resident groups in charge. However, the neighborhood pump is also susceptible to damage and breakdown. As the pump turned into an object of everyday use, repairs quickly became necessary. The neighborhood's communal coffer (*tabung*) didn't always suffice to pay for essential maintenance. I witnessed on several occasions how Arief was obliged to remind his constituents at community gatherings (*rapat RW*) of their financial and ethical responsibility. Arief justified the added economic burden on riverside dwellers as ethical work in the name of the future of the neighborhood.

Framing auto-construction as a form of religious labor is typical for Indonesian kampungs. Self-governed kampungs on the margins of urban centers can be traced back to colonial land governance practices (Cobban 1974) that strictly divided Dutch property and "native" land (see chapter 1). Economically autonomous but politically marginalized, kampungs developed sociocultural mechanisms to compensate for the absence of welfare structures and public infrastructure. While the Indonesian state has devised several biopolitically motivated schemes to improve living conditions in kampungs (Kusno 2000; Silver 2011), urban neighborhoods retain a certain degree of social autonomy. Jan Newberry (2008, 241) describes kampungs as community forms "reproduced through governance across various regimes but also through daily exchanges and support between inhabitants." The case of Arief underlines the tremendous role that personal labor plays in creating and maintaining a viable environment and working infrastructures. In a social-entrepreneurial spirit, Arief has urged his community to take matters into its own hands. But in the absence of long-term public investment in the area's infrastructure (see chapter 2), even his collectively owned pump, protected from water by a massive concrete hedge, simply loops water, effort, and labor in short circuits.

In fact, the government evaluates local labor, such as Arief's, as a positive example of self-help and progress because it shows that residents acknowledge the importance of working toward collective well-being by caring for the environment. At a certain point during my research, after news broke that the city government would expedite a delayed flood control project (namely, the Dutch-designed polder discussed in chapter 5), Arief was jubilant. In the neighborhood, there was a feeling of contentment and hope in the air. Residents felt assured that *rob* would soon lessen or disappear and that their efforts had played a significant role in this. They had made it through the rainy season without major flooding; seasonal and economic rhythms proved successfully synced, as the kampung infrastructure was raised just in time to resist rising water levels. At a neighborhood meeting presided by Arief that aimed to be ready for the mayor's visit, the lurah held a zealous appeal. Of note is that the lurah, a state representative, rarely attended such meetings, but, as mentioned previously, Arief's community had shown exemplary commitment and willingness to cooperate with the local government. The community would therefore play an important role in the mayor's upcoming river inspection, the lurah announced.

The lurah's promissory speech reflects the importance of the large-scale water management project for the subdistrict. He also echoes an official discourse that encourages communities to be "united" and "interested in improving the environment." Although several other subdistricts would benefit from the antiflooding project, he makes it sound as if they, in particular, had been rewarded for changing their "old habits." I suggest the contentment and jubilance show that residents felt they had become the rightful targets—the ethical substance—of governance projects, such as the polder.

On the day of the mayoral visit, Arief had a stroke. His face was disfigured; one side of his mouth had dropped. Because strokes are quite common among middle-aged adults in Indonesia, Arief was immediately aware of his affliction, but he continued the inspection along with other public figures of his neighborhood. People commented that he looked and talked funny. Weeks after the incident, I accompanied him to a weekly checkup at the hospital. He had been diagnosed with a stroke due to exhaustion (*kecapaian*) and was put on strong medication. Despite admitting his extreme tiredness, he explained that he would continue to "serve the community so that his life would not be useless" (*mengabdi supaya hidup saya tidak sia-sia*). In the waiting room of the hospital, he told me that he remained hopeful that by forming a cooperative pumping community and working hard, they would be able to "combat poverty" (*melawan kemiskinan*). After the checkup, we had *nasi pecel* (a cheap rice dish served with water spinach and peanut sauce) in the hospital canteen before he got into the long line for his medication at the pharmacy.

While Arief stressed the possibility of improving existing infrastructure by putting responsibility and resources in the hands of residents, the infrastructures supporting this arrangement also constantly verged on breakdown. When the limits of Arief's bodily capabilities and endurance become evident, the lurah announces that investments would soon flow into their neighborhood. While his personal efforts of improving the neighborhood are lauded by the kelurahan, his physical suffering remains unnoticed. The government seems unaware of how much strain is constantly being placed upon him: the continuously crumbling kampung infrastructure, the immediate needs of residents, and the exigencies of living in a toxic environment.

Act III: Worlds of Repair / Bric-a-Brac II

One afternoon, on my way home from city hall, an acquaintance waved me over to have a chat. What ensued was a half-serious conversation about work and correct attire. He mocked me for saying that I just returned from *kerja* (work) because I wasn't visibly sweating (*tidak keringatan*)—his hands gesturing toward a burst of water streaming down his face. He corrected me: "you're coming home from *kantoran*," the office or the office world. He then ridiculed my leather sandals, a fashion choice he considered *sombong* (arrogant). Embarrassed and feeling increasingly out of place, a feeling that I knew all too well that related to my obvious privileges as a white male foreigner, I wondered what connected these worlds: the air-conditioned world of the bureaucrat and the sweaty world of the coastal kampung resident. In the age of *rob*, they both inhabit an aftermath. They both belabor the obvious: decaying infrastructures built in colonial times and inherited by an independent Indonesia. Both are therefore concerned with maintenance, but the effects of their efforts never last long enough—inevitably another round of breakdown will obfuscate their partial successes at sustaining stability. There are no victors in the world of repair, as Jackson (2014, 227) argues: this world "tends to disappear altogether, or at best is relegated to the mostly neglected story of people working to fit such artifacts to the sticky realities of field-level practices." In a sense, kampung residents and government planners "inhabit a time that never arrives and the half project that never resolves, never completes, that changes into a frozen breakdown, yet secretes crime and half-solutions in the meantime" (Povinelli et al. 2014). When we look "past the Romantic tradition of inspired cataclysmic becoming and inside of its ruin," argue Povinelli et al. (2014), we see a "tsunami of quasi-events" where "potentiality is the refuge not of the hopeful but of the concretely ordinary and pragmatically banal."

As I got ready to take leave from this awkward exchange, Arief walked by. Actually, he appeared to be sauntering. Wearing his work outfit, he looked clean and composed, his white-collar shirt tucked into his long, ironed pants. He told me that he just got home from the baseball factory and had to check on the water gate (*pintu air*). "*Rob* has already started. I need to close it so that the water doesn't enter later." I went with him to the water gate, where the small conduit (*saluran*) emerged from underneath the embankment. The water gate has been broken for a long time, but residents fixed it with a little help from the government. Now the water gate was operated via a submerged hole and a piece of pipe that acted as a homemade stopper. Arief grabbed the pipe and pushed one end into the dark liquid. He found the hole easily, the gate closed.

That same night, I returned to the *dewan* and ended up hanging out with Arief and his clients and friends. As Arief cooked some *mie rebus* (boiled noodles) for me, it started raining. The air immediately cooled down. People kept flocking to the *dewan*; there is no better time to mingle with neighbors than during a rainfall—the air feels substantially cleaner. As I waited for my food, in walked an acquaintance, a young man called Imam, who used to work at the Chinese-owned laundromat on the main street. I had been wondering what he was up to since I hadn't seen him lately. I had started to worry about him, especially when I heard that his daughter had badly hurt her foot in a motorcycle accident. How could he care for her without a job? As I was eating my dinner, I noticed that people didn't treat Imam with much respect. They seemed to ignore him and they cracked jokes when he showed up. But Imam was a good sport; he laughed it off and started to massage the shoulders of the old man I was speaking with prior to his arrival.

The old man had been telling me about his new business ideas. He ran a chicken farm and cut (*jagal*) throats all day—up to a hundred times. Now, the government had announced plans to evict him and other vendors. The whole market would be relocated and his business would have to adapt. His fellow vendors supported each other despite being competitors; it was a moral obligation to have one another's backs. The canteen run by Ariel and Arief was also a fairly tenuous business model. Arief was more or less waiting for the government to order them to pack up and leave. All along the embankment, structures, some of which are shacks inhabited by whole families, had been erected. According to the plans of the government, they would have to close one day. Arief assured me that he wouldn't protest if the government eventually ordered the demolition of their food stand. About five years ago, the government had announced that normalization works would wipe the riverbanks clean of built structures. According to the plan, the river was indeed widened and embanked, but the spatial regime installed on Banger's southern banks never quite thrived in Kemijen. North of the bridge that separated Arief's neighborhood from the South, "illegally" built structures remained untouched. I asked the nighthawks how this was possible, and they answered that the government was afraid of *preman* (gangsters). What could *preman* do against evictions, I insisted. "The government is afraid they would steal construction material," one man answered. Whether this was true or not, normalization did nevertheless intersect with residential projects. In the case of Arief's *warung,* it created a transitional space where people could meet to make plans together when faced with motions that they can't control alone, as in the case of the chicken vendor. Here, people could find a way to use a government plan as an instrument of action (Baxstrom 2012). In Semarang, as elsewhere, an eviction often means the establishment of a new order, in terms of both time and space. In Kemijen, however, evictions didn't establish a new order on top of the old one; instead, it allowed residents to inhabit both. When the government eventually makes good on its eviction plans, Arief won't stand in the way of it because in his view it will open up new opportunities. But for the present, the riverbank, where he caters to locals and his anthropologist friend, is his source of livelihood.

Act IV: In-Filling

While waiting to attend an engineer's presentation on drainage in Margorejo, a northern neighborhood of Kemijen, I met the area's neighborhood chief, Jusuf. We sat down in front of the neighborhood hall (*balai RW*) that adjoins a sports field. Jusuf proudly told me that before he took office, the field had been a filthy water hole teeming with mosquitos. He had since seen to it being paved over. Although he had acted in the interest of his community, in an effort to create additional public space, he had simultaneously reduced the catchment area, thereby increasing the likelihood of flooding.

Impromptu interviews such as this seemed to be a welcome opportunity to kill time in Kemijen. The stretched present worked to separate people from a different kind of future, and sometimes this present seemed to contain an awful lot of time. On a different day, I visited Edo, another eager interlocutor. When I arrived at his home, I was struck by the way his house served as a shockingly accurate example of how accelerated environmental decline translates into economic stagnation and vice versa. Edo moved to Kemijen four years ago, in search of a job. At the time of the interview, he worked at a hotel across town. His whole family, including his two teenage sons, must work to make ends meet. Still, they didn't have enough money to be able to undertake a "real" renovation of the house. The house, which belonged to his dying father-in-law, who was spitting mucus and blood into a bucket in the damp back of the room while I interviewed his son, stays above street level only by a nose. Its doors and walls are cut off, like amputated limbs. The rods and roof tiles hang low over your head, and maintaining a semi-crouch is necessary in order to prevent a painful encounter with a door frame. The house was teeming with mosquitoes. While showing me his scarred underarms, Edo complained about getting bit all the time. I offered him some of my *minyak sereh* (lemongrass oil), which helps deter these pesky beasts for about twenty minutes, and he quickly applied it to his legs. Then he took another shot and applied it to the same spot, over and over. He wouldn't stop; he even put some under his T-shirt, where it obviously didn't have a preventive effect. He smiled and said that it felt hot (*panas*) on his skin. I couldn't help feeling as if the interview was worthwhile for him, like a welcome distraction. As I was leaving, he asked me to return soon.

Outside the house, I glimpsed another window into Edo's relation with time. The slits above the low doorframe were filled with empty cigarette boxes. As used items that represented time gone or spent—the breaks during demeaning room-service shifts at a local hotel, the evenings playing chess with neighbors, the long nights of scratching mosquito bites—they were now empty decorations of a house outrun by time.

Act V: (No) Escape

With no warning, Imam disappeared one day. He stopped coming to the *dewan* to massage customer's shoulders. Arief, who not only screened the water surface but also kept track of the comings and goings of residents, informed me that Imam was in hiding. Nobody knew where he had gone, but Arief heard through the grapevine that he took refuge with relatives living outside town. Imam had apparently amassed significant gambling debts. Weeks earlier, in the course of our only conversation, I had learned about Imam's difficult situation. After finishing elementary school, he lived mostly on the street, making money as a busker. His informal employment in a factory didn't earn him enough income to feed his daughter and himself—his wife left them a few years ago. As a result, Imam was often hustling for extra money.

> I'm all right! Can't change how it is. If I'm hungry, I'm hungry . . . What can we do to get food? It's just as it is. When you're sad, things are difficult. When you're happy, it's fine. People can't tell. That's how I was raised. . . . Perhaps my dad taught me. Whether he liked his meal or not, people couldn't tell. It's really sad, if your life means being burdened by debt, but you stay strong. You carry this load on your own [*dipikul sendiri*]. Holding out and being patient is a skill. That's key in life. Sure, today I really don't have anything to eat, but I resist, I have to fight. I'm often hungry. Really often! But I have something you could call a talent: I'm good at massaging. So, I go up to people and say, "Please, sir, I don't have any money." "All right, all right." I'm already used to giving massages to friends. Then we go to the friend's place and eat together. One year, my kid rarely ate, because I couldn't always give her school money. Because I wasn't working at that point. I was relying only on giving massages. At first, I just did it sometimes, but people actually felt better afterward, you know.

It was as if Imam didn't want to burden anybody with his problems. In the interview I conducted in his room, his voice constantly vacillated between sadness and acceptance. You wanted him to see a brighter future, but, really, was there one? Imam had inherited problems that he seemed unable to handle. Since his parents had passed away, he lived with relatives and a few orphaned children in an unfinished house. He shared a room with his daughter. The rough floor was covered by a stained and worn mattress, the only item of furniture. When the house became uninhabitable due to land subsidence, his father had to renovate because the roof tiles were already low. Imam's father took out a loan because he wanted his family to be safe. Imam remembered:

> The roof part was increased. The roof first. The walls were raised a little. At that point all the family members were gathered here. Like that. Back then. My older brother already had a child, but he passed away. So, we had to make the house bigger, whether we wanted to or not. So, my older brother died and the children he left behind all moved here.

More and more family members moved into the incomplete house built on debt. Interestingly, exposure to flooding used to provide a kind of social membership through which Imam's family could relate to other inhabitants. According to Imam, water doesn't differentiate between the haves and the have-nots. The shared inability to escape across social difference often created a common denominator of social belonging. In fact, I observed that monsoon floods to this day are cheerfully celebrated. *Rob*, however, does not become a bonding experience in the same way. Rather, it occasions differential investments in flood protection. Imam's abandonment is thus marked by the arrival of *rob*, as it introduced privatized flood management.

> Whether you want to or not, all citizens are protecting themselves [*talud-talud sendiri*]. . . . Every citizen owns a pump now. They pump the water out on their own. It wasn't like that back then. If it happened, it happened to all [*kalau kena, kena semua*]. They were hit by water every day, *rob* happened every day. . . . The government made the walls, and then citizens began using pumps. That became an asset; you just needed to hook it up. I still remember! It was difficult before we had the riverbanks. Because the river was as high as the street back then. You couldn't pump the water out! It entered and exited, when it pleased. Then, it was widened; the river was walled. So, the citizens took initiative to build and buy a pump. All of them have one, all of them bought their own pump. So, it's pretty nice now, much better than in the past.

"It's pretty nice now," Imam said a few weeks before he disappeared. His observations and experience of privatized flood protection in the wake of normalization show that the kampung, as a world of repair, can squeeze people out. Existence built on borrowed time has an expiration date. As there seems to be no escape from the present, Imam longs for a past when "there was always floods. We walked in floodwater, we slept in water [*tidur sama air*]. We were used to it. It was like that for years."

Act VI: Some Strategies

Strategizing is never a straightforward process; it involves guesswork, speculation, and patience as well. Once, while getting a haircut, I asked the shop manager, Agus, why he and his wife, Dina, chose to live in the area. He previously worked at a machine shop (*bengkel*) and as a house constructor (*kontraktor*). Now, he manages the couple's salon, while Dina cuts hair. When he is not handling customers, he entertains his youngest child. Currently, they hold a one-year lease for the salon, but they are looking to purchase their own business space out of fear of unstable rents. Despite *rob* and the area's stained reputation, they are willing to stay here. If they bought property, it would have to be located in the vicinity. Agus explained, "It's more strategic. I can go anywhere, and it's close to the [bus] terminal. It's a strategic location; I can ride the motorbike anywhere. . . . Family members live close by; my brother-in-law lives on the other side of the river. We often see each other."

Agus and Dina are also compelled to live in Kemijen because they see opportunities for investment. Dina's family has lived in the neighborhood for a long time, and she has created a client base. They have made up their minds, coupling economic strategizing and endurance. Notably, the income generated by the salon was funneled directly into renovations. I recorded the following conversation between the three of us at their hair salon.

> AGUS: We only renovated once, because we built the house in 2006. It was raised in 2011. The reason was flooding. In the past, it was flooded every day, before the renovations [*Dina cannot suppress an embarrassed chuckle*]. Whether it rains or there's *rob*, those floors absorb [*peresapan*] the water, right? We spent around 16 *juta* [approx. USD 1,200] for all renovations: the restroom, the kitchen, and the floors.
> DINA: Including drying the house [*jemuran*].
> AGUS: Thank God, we're flood-free now. For another five years . . . [*he laughs*].

DINA: Perhaps up to fifteen years!

AGUS: The rear part of my house is about seventy centimeters higher than the house front. The porch of my house is as high as the guest room [*ruang tamu*], but the second room is thirty centimeters higher. The kitchen is higher by about . . .

DINA: One meter.

AGUS: Well, more like seventy centimeters. So, the back is higher than the front.

LUKAS: It's seventy centimeters higher than sea level?

AGUS: No! It's seventy centimeters higher than the front.

LUKAS: And if you had to compare it with the level of the river?

AGUS: My house still wins, of course!

The conversation turned into a debate between the spouses about the height of their house in relation to the street. While they didn't share the same feelings about the level and duration of safety that the height provides, they agreed that they had to adjust to the street level. The river is not the culprit but the street: its height determines whether their abode is at risk. The house floors could soak up ground water. The couple demonstrates an endurance that is different from all other forms of endurance I have discussed thus far. They cloak this endurance in economic survival. This becomes particularly evident when they frame themselves as obeying local regulations. For example, Dina previously ran her salon by the river but had to vacate the spot when the government ordered residents to clear the riverbanks in the context of normalization. She didn't complain when others, in turn, began to build houses and kiosks on the vacant land: "In the end, new buildings were erected on the street that I had to clear. That space has in fact evolved. People considered moving there because the [planned] constructions were delayed. So, in the end they just set up shop there themselves. We can't blame them either. Like . . . when people are starving, we can't do much. It's because of the delay [of normalization]." She then called this form of endurance *urusan perut,* which translates into "living from hand to mouth," and separated out individuals with these unmet primary needs as requiring more urgent arrangements. She contrasts *urusan perut* with their own endurance, a practice of strategy responding to local economic shifts and competition.

A Photographic Interlude: "Baroque" Architectures of Time

The following photo-ethnographic tour of a residential street of Kemijen pays attention to the architectural aspects of floating as a response to flooding. Houses, as primary infrastructure, maintain an intimate relationship with the time spans of public infrastructure, weather patterns, and budgetary cycles. They are the vessels that allow residents to survive in the region. They represent strategic plans, that is, hopeful adjustments to infrastructural changes whose projected outcome—flood safety—exists only as a "formal gesture to the linear ideal of 'the plan' itself" (Baxstrom 2012, 138). As *rob* continues, infrastructural development plans shape and create the "material conditions of the more individualized and singular modes of planning exercised by individual city residents" (138). As such, I call these architectures of time "baroque," as they demonstrate a "disposition," in Baxstrom's words, that urbanites develop in the face of a seemingly incongruous present. Instead of orienting themselves to one futurity at a time, set out by a linear plan, their strategies represent a flexible mode of living in the present. In the following section, eight photos of house entrances and facades show the (mostly serial) adjustments of residents to changing environmental norms that often result from overlapping plans that coexist in the present (Figures 6–13). These adjustments are not coordinated. Rather, they are the result of a wide variety of factors, such as financial ability, access to building materials, and level of community integration, all of which underline the unequal socioeconomic context of tidal flooding. The photographs also quite simply serve the purpose of familiarizing the reader with the aesthetics of a coastal and mainly poor Semarang neighborhood.[4]

Postlude

This chapter proceeded from descriptions of downstream residents' struggles with the consequences of *rob* and moved on to an appreciation of infrastructural projects that play a role in preserving yet "hardening" the northern wetlands of Semarang. It interrogated the local experience of this hardness, developing the claim that residents inherited a landscape that was not of their choosing but one that they nevertheless have to endure. I let the meaning of hardness derive from a resonance between personal observations and representations of residents. In doing so, I follow Michael Jackson and Albert Piette (2015, 5), who have argued in favor of an "existential anthropology" that focuses on the microsocial in order to restore to our worldview "a sense of the small and tangible things that make life viable and negotiable despite the forces that elude our comprehension and control." My gaze thus shifted from the macro to the micro in order to capture the small things and events that allow residents to inhabit the hardness of life in Semarang's delta. Actions such as continual renovations or Imam offering massage to Arief's clients in exchange for food speak to the limits of local resources and strategies for self-help, but they also provide testimony to the will to endure. However, Imam is exhausted and frustrated by his daily struggle. Likewise, Deni works every night on the roadside despite a meager turnout of customers. His constant head-shaking and sighing betray a lack of avenues—he can't run his *warung* by the river, a preferred business location, because the government enforces inspection regulations. He bemoans a present with limited possibility. A few hundred meters north, police enforcement of spatial regulations is spottier, which allows Arief's family to generate extra income to help them endure regular exposure to floods. He maintains the space as if it were his official duty. His volunteer labor is rewarded in kind by the community and in lenience and acclaim by the subdistrict government. The consequences of his work are bodily: a somnambulant existence, a stroke. In the same neighborhood, the entrance of a sunken house is adorned with empty cigarette boxes whose contents once brought welcome distraction from a deteriorating environment teeming with mosquitoes. Elsewhere, relief from these bloodsuckers requires filling in the swampy hole with concrete to eradicate breeding grounds, while also problematically reducing crucial water retention space.

  The hardness exposed in this chapter can be further explored via the concept of infrastructural violence. I conceive of infrastructural violence as stresses regularly experienced by individuals as a result of infrastructural systems. Akhil Gupta (2012, 20) has argued that "structural violence is constant rather than episodic, and far from disrupting actors' understandings of their social words, it provides them with a particular kind of situated knowledge with its own epistemic certainties." Flood infrastructure, a composite biotechnical assemblage of technology, cement,

Figure 6. The houses in this *gang* (street) use an open sewer system, which means that in the event of flooding (either tidal or seasonal), river water combines with sewage—a toxic brew that people want to keep outside at high costs. This elevated and thus flood-safe house entrance is richly decorated with papaya, chili, and other plants, which provide shade and a pleasant hangout and work space. The slightly elevated plants are kept out of the reach of small floods. Photograph by the author, 2015.

Figure 7. This recently finished house was uninhabited during the period of my fieldwork. The house's remarkable height suggests that the owner considered the house a long-term investment, as it will outlast many years, perhaps decades, of land subsidence. In contrast to the previous house, the stairs are ornamented with differently colored and shaped tiles, which further suggests a relatively wealthy owner. As it turned out, the owner had in fact built this house for a close relative. However, the relative never materialized and the house remained uninhabited. The owner couldn't find a suitable tenant for such a palace-like building. Photograph by the author, 2015.

Figure 8. This house was finished during my fieldwork. A young family moved in. Its design—both coloring and level of completion—was extraordinary given the mostly fragmentary and evolving nature of most houses of the alley. When I passed by this house accompanied by my research participant Sumarmo, he joked that the owner of this house clearly didn't trust the government (*tidak percaya pemerintah*). Despite politicians' promises to stop flooding in the area, the house was built with the current flooding trend and exorbitant land subsidence rates in mind. Another aspect is noteworthy. The owners weren't well connected in the region. Unable to rely on networks of protection, they decided to equip their house with special security measures: the gate and grid. Photograph by the author, 2015.

Figure 9. Motorcycles can be kept at a safe distance from flood water at the entrance to this house. Should the inside fill up with water, bikes can also be parked outside on a raised plateau. The wooden door shows traces of past flooding, as the lower end shows a lighter shade of brown. The door predates the current iteration of this house. Photograph by the author, 2015.

Figure 10. The strategy of this resident consisted of building up. Often, whole families slept on the second floor, as this space was never flooded. The lower part of the house is highly exposed to flood water. However, the household benefited from being located closer to the main street, from where the rest of the alley slanted down. Photograph by the author, 2015.

Figure 11. The ramp in front of this house entrance allows residents to shelter their motorbike(s) inside from rain, floods, and thieves. The brick wall is unfinished. The members of the household probably decided to raise the roof recently. The next step will consist of adjusting the floor height. Photograph by the author, 2015.

Figure 12. Without stairs, this ramped house entrance allows its seated owner to easily access the inside. The concrete ramp itself shows the different adjustments that have been made over time: its layered segments show successive periods of construction. Each time the gang is lifted, the ramp must be slightly adjusted to remain straight. Photograph by the author, 2015.

Figure 13. The fresh concrete used for this ramp stands in contrast to the moldy street-facing walls. The house floors had to be raised several times, which decreased the ceiling height. Today, the tiled roof hangs low over the heads of the members of this household, and adults must duck their heads to enter. The subsiding ground has further destabilized the foundations of this house, which explains the slanted roof, propped up on additional rods. Photograph by the author, 2015.

and residential interferences, makes it impossible to identify a single person or organization that caused flooding, that is, committed the violent act. Rather, the living environment itself exposes people to risks and violence. As a result, people consider themselves victims of arbitrary processes. But, as Gupta contends, "such arbitrariness is not itself arbitrary; rather, it is systematically produced by the very mechanisms that are meant to ameliorate social suffering" (24), such as regular street raising or house renovations. Contemplating the normalizing impacts of structural violence, I'm reminded of Djawahir's poem, wherein he mourns the absence of water flow as a metaphor for the absence of social change, or its very possibility. *Rob* itself has replaced this potential of social transformation by becoming the pulsating vein of everyday economic life. A "normal" river floods periodically: a sign of benevolent circulation and change. A stagnant river can only swell and seep, with corrosive effects on the material structures of everyday existence. Water seeps through the cracks of floors, abrades water banks, and remains trapped in kampung streets and people's houses. Residents have organized politically and socially around this phenomenon; they have acquired pumps and formed pumping associations to restore water flow, often by taking on crucial governmental tasks. Yet this structure of feeling disavows the potential of full breakdown. Rather, *rob* is perceived as opening up possibilities for linking with larger structures and plans through community empowerment and programs for economic progress.

In this chapter, I also demonstrated how residents' endurance is textured by prescriptive state programs that encourage participation and self-management. *Rob* attracts new infrastructure and community development projects that allow residents to generate the resources to continue to endure in the present. But these programs never fix the ruined riverbanks; they just patch them up. The result of the coming and going of projects is that floods and the ensuing suffering are framed as quasi-events that do not disrupt the order of things. Invisible forms of suffering—exhaustion and sickness—are a product of enduring in the everlasting present.

To close this chapter, I return to the comments made by the Dutch mangrove expert regarding Semarang's coastal infrastructure. On the face of it, her assessment of a solidified urban environment perfectly captures the "hard" reality of riverbank dwellers. But as I have illustrated, the situation is far more complex than that. The hardness of infrastructure—in its material form, and its imagined promise—does not translate into certainties and safety from flooding. Riverbanks are alternatively seeping or breaking. The lives of my interlocutors expose the fragility and unreliability of infrastructure, as subsiding ground drowns and swallows houses in ever-shortening intervals. In this catch-22, they have two equally unsatisfying options. They can wait for government improvement projects, like the polder, and do nothing in the meantime, putting them at risk of gradually losing their homes to the rising tide. Or they can actively inhabit an environment that is at the mercy of tenuous and fragmentary interventions. This chronic engagement with risk consists of preventing hardship by repairing infrastructure. At the same time, ruptures produced by breaking infrastructures are folded into the experience of riverside life. Yet existences built on borrowed time have an expiration date, and repair can only stretch the present for so long.

4

FIGURING
ENVIRONMENTAL GOVERNANCE AND THE POLITICAL AFFORDANCES OF INFRASTRUCTURE

Nature's Tenants

"We own our houses, but we pay a rent to nature" (*sewa pada alam*). I heard this joke among residents of Kemijen many times: chatting with batik vendors, eavesdropping at city hall, and in interviews with close informants. The joke expresses an awareness that ownership and infrastructural investments have a negligible effect on the delta inhabitants' precarious cohabitation with water. Although the municipality mobilizes funds each year to elevate streets in Kemijen, residents still have to cover the costs to raise their own houses on a regular basis. As people say, this creates a mad "competition" (*lomba-lomba*), where the efforts of one resident can negatively impact the circumstances of another. The higher the streets on either side of the river—which also function as embankments—the more difficult it becomes to drain the neighborhood of rain, waste, and tidal water. If houses don't stretch themselves upward, then gravity and water combine to become ever-more destructive.

While the common joke of paying rent to nature is a humorous take on environmental risk and uneven levels of adaptation, it also exhibits both an attachment to place and an accompanying stoic pride in the ability to endure. As I demonstrated in the previous chapters, coastal residents continue to defiantly invest in meaningful ties to a volatile environment where they have endured floods, disease, and economic crises for decades. Paying rent to nature is not the same as paying rent to a landlord. It is an investment in a valued relationship with place. As one resident told me, "Be it banjir [flooding due to rain] or *rob,* people feel comfortable here." Exposure to flood hazards is an acceptable trade-off in view of the area's proximity to major employers (mostly foreign-owned factories) and services (education, public transportation, and municipal administration). Notably, despite the area's strategic location, property prices have not increased significantly in the past ten years due to flood hazards.

Taking individual pride in enduring floods, however, masks a strong dependence on public infrastructure investments, such as riverbanks, streets, dams, and flood gates, which demand continual maintenance. These irregular gestures of repair and construction reproduce and structure the kampung—along with its hydrological quirks. As I showed in the previous chapter, repair carried out by residents and government-conducted repairs play a crucial role in preventing floods. In the context of state decentralization, residents have become increasingly important partners in carrying out public repair projects. These projects produce specific publics that confront the contingencies of *rob* in diverse ways. In order to tease out this process, I focus on residents who have begun to seek out a more active role in coordinating infrastructure provision. These residents argue that the missing link in flood protection—besides timely riverbank and road repairs—is participatory planning and development. They have come to use state-led infrastructural programs as an arena and discursive platform for this bottom-up "figuring." In what follows, I explore how new government-sponsored civil participation programs transform local conduct and how acts of coordinating bottom-up responses to *rob* in turn affect the social order in Kemijen.

Balancing Act

One morning, I visited the subdistrict office to follow up on a data request I had placed the previous week. I was curious to know how many residential infrastructure improvement projects the government had approved in recent years. Repair proposals are submitted to the lurah, the government-appointed official at the head of the kelurahan, for review. The kelurahan selects fundable projects and forwards them to the *kecamatan,* the district government, for approval. I wanted to know what kinds of projects had made the cut. As I waited for the treasurer to photocopy the handwritten records, I was given an excellent opportunity to more carefully take in my surroundings. Kemijen's kelurahan office, built in 1980, is a modest version of a typical Indonesian local governance post. The office furniture looked worn and due for replacement. Visitors were expected to sit on wobbly blue plastic chairs, as were the lower-ranking officials in the lurah's employ. As I was waiting, the community organizer Wahyu walked in. The moment he noticed me, he smiled and approached. After shaking hands, I asked him what he was doing at the kelurahan. He casually informed me that he had come to discuss an issue with the lurah. He was apparently in a rush, so he promptly excused himself and walked in the direction of the lurah's office, where he was immediately received. For the next ten minutes, they spoke about something in what seemed a relaxed fashion while the door was kept demonstratively open. Wahyu had recently been elected director of Kemijen's "core planning and marketing team" (Tim Inti Perencanaan dan Pemasaran), a group tasked to design a bottom-up infrastructural renewal program financed by the World Bank. Wahyu is also a member of a citywide nongovernmental network that monitored water policy making and public spending on kampung water infrastructure.

 Later that day at Wahyu's house, I learned that his visit with the lurah had to do with an alternative antiflooding policy program that another nongovernmental organization, Pattiro, was concocting. Wahyu collaborated with Pattiro to collect residential input for the program. Therefore, he thought it necessary to inform the lurah that his staff had neglected to invite certain residents to the organization's next public consultation. Excluding these residents from their meeting, he feared, would prove problematic because it would create the impression that the kelurahan cherry-picked its partners. Wahyu knew that when residents felt left out, they might boycott or block important programs. He wrapped up their conversation with a respectful reminder: "time is moving on, and people continue to have needs."

 Witnessing this encounter at the kelurahan led me to read the subdistrict's spending records with a grain of salt. While I was out for "hard" facts that could explain the infrastructural situation of the neighborhood, I observed in situ the politics of infrastructural development.

This politics is characterized foremost by relations of mutual benefit between civil actors and the local government. Note also that the encounter between Wahyu and the lurah took place at the subdistrict office—a formal space. Their conversation was in theory audible to the whole staff, including visitors, which suggests adherence to ideologies of transparency and democracy—the cornerstones of Indonesia's decentralization policy. Furthermore, Wahyu invoked the rakyat, or people—a powerful concept in Indonesian development politics and national history—to assure an implementation style that was socially inclusive.

This chapter revolves around the question of why this "ordinary" resident was granted exclusive time with the lurah in order to remind him of his official duties. Wahyu's audience with the lurah demonstrates something crucial about state–society relations in Indonesia in times of infrastructural breakdown: good relations between community organizers and the local administration are mutually beneficial. The timeframe in which Wahyu located the situation of the "people" is also important; it applied to the lurah's window of opportunity to do a good job while holding office as well as residents' present and ongoing problems with tidal flooding. If they wanted to make progress on achieving flood safety while he was in office, they ought to work together and be as inclusive as possible. As his message to the lurah shows, Wahyu claims to be speaking on behalf of the rakyat.[1] Whether or not the lurah agrees with Wahyu's view of the situation, he has a vested interest in cooperating. In fact, the lurah's success at governing would be measured by how many development projects (to reduce poverty, improve well-being, maintain local peace, upgrade infrastructure, etc.) he was able to implement.

While it is important to differentiate between public and private efforts to protect property from floods, they become entangled in the preservation of the neighborhood. On the one hand, the government is responsible for raising streets and public facilities, and supplying and maintaining certain parts of drainage infrastructure, such as pumps, pipes, and gutters. On the other hand, residents need to continually invest in the foundations and roofs of their houses and in other general household repairs. Public investments translate into tangible consequences for residents, as repaired water infrastructure to some extent hedges the risk of water-damaged individual property and assets. As such, public and private infrastructural investments exist in a delicate balance, often conditioning but also undermining and offsetting each other.

Rob needs to be viewed as a phenomenon that is produced by this very balancing act. *Meluap* (overflow) does not mean banjir (flooded) for all, as only houses that lie beneath street level at the time of overflow are affected. In other words, even when the river spills water into the neighborhood, the levels of some houses might temporarily exceed the level of the water (perhaps only by a nose). However, given the unpredictable pace and spatial irregularities of land subsidence and the differential life spans of infrastructures, that might not be the case in the following year.

While I focused on the individual experience of flooding in the previous chapter, I now describe collective responses to flooding. Uneven levels of flood safety result in tactical water-control practices that put neighborhood groups in competition with one another for government support as well as funds from private actors, such as businesses and universities. Here the success of generating funds has come to depend on how one positions oneself in relation to infrastructure and kampung development. The chapter is titled "Figuring" because now that the area figures more prominently in government plans, residents have to try to figure out what roles they can play in the increasingly participatory governance of flooding. Knowing what is in store for one's area in terms of flooding and potential infrastructural responses makes the future a less slippery thing. It also affords political and social capital that can turn out quite useful in navigating the modalities of decentralized governance in contemporary Indonesia.

At a theoretical level, this chapter speaks to the affordances of river infrastructure in light of political transitions. I consider new conversations between the state and society as inflected by infrastructural life spans and historical changes in Indonesia's politics of governance. I illustrate how individuals develop sensibilities to water governance as a result of shifting environmental and ideological circumstances. Furthermore, I compare the development projects of different neighborhood groups (those identifying as activist as well as nonactivist) and demonstrate how they lobby for increased governmental support by posing as subjects concerned with the environment and local well-being. Based on these findings, I query the role that water and infrastructure play in the local governance of tidal flooding. Similar to Emma Colven (2020, 314), I show how a heightened visibility of water infrastructure can be "leveraged or produced by different groups." *Rob* opens up spaces for the development of grassroots politics in North Semarang. The temporalities and material contingencies of water and infrastructure, as well as their in/visibilities, need to be taken into account in order to better understand these grassroots politics.

In the sphere of Kemijen grassroots politics, strange new assemblages have emerged: for example, alliances between social activists and the kelurahan, the subdistrict administration. While the latter used to be a governmental instrument of social control, activists and the kelurahan now closely collaborate to acquire funds for infrastructure improvement through state development grants (like PLPBK) and the municipal participatory development scheme called "Musrenbang."[2] By thinking through relationships between grassroots politics, models of democratic governance, and materiality, this chapter examines the influence of state ideology on flood governance. Though ostensibly democratic, development projects often subsidize more than just infrastructural adaptation to flooding insofar as they tend to reproduce existing power structures. Decentralization and the effects of a neoliberal government agenda (promoting competition for funds) result in residents having to constantly hustle for project money at the municipal and provincial level; this in turn reaffirms and deepens the uneven landscape of the coastal kampung. While government funds or development grants are mostly used to level-up streets and kampung infrastructure (peninggian), little money is provided for community development, education, and internal economic growth. In building on activists' critique of peninggian, I contrast the concept of pembangunan, a powerful technocratic term that frames development as a process of awakening, building, and construction (Barker 2005), to peninggian. The widespread practice of the latter demonstrates that the provision of infrastructure has undergone a subtle change in meaning. Stacking pavement layer upon layer no longer suggests the coming of a glorious future. Unsurprisingly, this type of building is also not identified with nation building; instead, it suggests an endless corridor to progress, where northern residents can be parked in the everlasting present.

The Grassroots Structure, Political Transition, and Developmentalism

In contemporary Indonesia, the grassroots often function as an anchor for development projects. In the framework of projects, residents take on not only the roles of beneficiaries but also those of organizers, contractors, workforce, or facilitators, that is, "active forces supporting development" (Hadiz 2004, 700). As Tania Murray Li (2015, 84) has shown, a successful project is, foremost, one that "generates a flow of revenue, and more projects." In line with her observation that policy is not the only tool of governance in Indonesia, and a weak one at that, the project has become a privileged mode of governing conduct and the use of space in Semarang's North. The actors that the Indonesian president has called the grassroots (*akar rumput*; see beginning of the chapter) play an important role in the formation and implementation of projects. Through providing historical context to my discussion of *rob* governance, I analyze projects as a form of governing *rob*. The purpose of this section is to show the new relationships between the kampung and the state in Indonesia and to put Kemijen's own "*rob* activism" in context.

In Indonesia, the urban grassroots are generally considered as developing and emanating from kampungs, that is, the living spaces of the lower classes. Following Yoshi Fajar Kresno Murti (2015, 48), I see the concept of the kampung operating on at least two levels in grassroots politics. First, it functions as a space of identity formation, and second, it forms the basis on which activists can negotiate and advocate the needs of communities. The grassroots are thus coupled with the kampung and its administrative structures. To unearth the Indonesian grassroots—a rather slippery concept often used to refer to the "ordinary people" that make up civil society—a few more words on Indonesia's political transition since 1998 are in order.[3]

During the New Order, the kelurahan was an important governmental tool of power and suppression. At that time, "the kampung was a unit of control" (Murti 2015, 45) under tight surveillance by the local administration installed by the central government. The kelurahan exercised a hegemonic force that promoted a uniform view of time that tended to "focus on the logic of developmentalism, integrating competing sources of knowledge and ways of narrating the past into a single national narrative of the Indonesian nation and its people" (45). The mid-1990s, however, sparked optimism among government critics (Tsing 2005, 22); pushed by foreign foundations, the regime was suddenly inclined to foster community-based resource management, as long as such projects aligned with the interests of the "nation."

After Indonesia's transition to democracy, community projects suddenly proliferated. Today, the kelurahan is officially conceptualized as a space of cooperation, one that acts as an interlocutor and development partner for residents, to which the latter can express their concerns and desires. The kelurahan, for example, carries out *sosialisasi* events on a regular basis to inform residents about development plans, regulations, and programs. The Suharto regime once employed *sosialisasi* in a more authoritarian way, and critique voiced by residents at *sosialisasi* events was not permitted. Today, *sosialisasi* is conceptualized as a forum for exchange and mutual support between residents and the government. Despite being designed to represent citizens' critiques and recommendations, Sheri Lynn Gibbings (2021) has shown that *sosialisasi* still largely follows the logic of educating residents and generating buy-in instead of constituting a veritable space of participatory planning.

In his analysis of contemporary Indonesian politics, Edward Aspinall (2013) notes a lack of the "powerful and permanent poles of attraction" that dominated the political landscape of postindependence (1945–66) and New Order (1966–98) Indonesia. He argues that the politico-ideological groups (*aliran*) that crystallized under Dutch rule and formed the basis of the New Order's clientelistic regime have lost their social magnetism in Indonesia's current "liberal" era. Furthermore, political and ideological fragmentation have weakened large-scale social movements and replaced them with a "silent majority."[4] When, in 2015, the newly inaugurated Jokowi government discontinued public subsidies on oil, causing gas prices to skyrocket, protesters took to the streets of major cities, including Semarang, but ultimately failed to form a mass movement.[5] In contrast, in 1998 President Suharto's decision to implement the International Monetary Fund (IMF) demand for fuel price increases was a significant factor contributing to the civil protests that forced him to step down. Again, in 2003, Megawati Sukarnoputri, the daughter of Indonesia's first president, Sukarno, shied away from plans to cut the oil subsidy after "large-scale demonstrations" (Roberts 2005).

This new, piecemeal way of protest is arguably symptomatic of a general lack of common ideals and shared imagined futures around which society might congregate. Decentralization—a set of policies concerning regional governance (Bunnell et al. 2013)—played a major role in this fragmentation of society. The weakening of a centralized state apparatus and devolution of power and financial autonomy to municipal governments and their districts undid former political constellations.[6] Decentralization coincided with cultural shifts; the introduction of neoliberal values arguably furthered the subsumption of everyday life under capitalism, with social life becoming tilted toward consumption, competition, and entrepreneurship. But more importantly, fragmentation has altered ways of being political. What this means in concrete terms for Indonesians is still largely unclear. For this reason, critical scholarship has been cautious to assess the result of Indonesia's political transition. Rather, it looks at how issues become politicized and actuate social formations. Aspinall (2013, 49) draws a pessimistic picture and suggests that while nepotistic networks of officials, politicians, and local bodies of authority came under pressure, thus opening up spaces for a democratic politics, clientelistic ways of divvying up wealth and opportunities nevertheless survived into the present. This had everything to do with decentralization and other structural reforms adopted in the wake of the "Asia Crisis." Taking the wind out of political movements' sails, decentralization undermined coalition building by introducing development programs based on competition (Li 2007). Today, building coalitions between mass-based organizations therefore fails "in the face of splintering at the grassroots" (Aspinall 2013, 35). Instead, grassroots movements at the level of neighborhoods find themselves atomized and forced to partner with the local administration.[7]

With the end of the New Order came hope and prospects. Who would be a part of the new social order forming on the horizon? After protests that led to the end of the regime, something new filled the power vacuum. In line with structural adjustment schemes of the World Bank, the nation's "communities" became instrumental in implementing development plans as direct recipients of state subsidies and donor money from international organizations. As Aspinall and others (Tsing 2005; Li 2015) have argued, the development project as a specific site of value production became the new glue of social ties; the *proyek* (project) was formative of assemblages of global and "local" actors, experts, recipients, and spaces. As Li (2015, 80) observes, enrolling "government officials, politicians, transnational donors, NGOs, scientists, and villagers . . . form[s] them as subjects, and engage[s] them in a particular set of practices." In the context of development projects, foreign NGOs encouraged the activation of the grassroots, arguing in a classically neoliberal vein that budgets and planning should be devolved (in part) to people on the ground.

Importantly, as a motor of national development, the *proyek* became a screen behind which the capitalistic elite, patrons, and former members of the New Order powerhouse could transform themselves into actors allied in specific ways and to variable extent with the "grassroots." In fact, what counted as a *proyek* addressing the grassroots often required creating recipients with specific customs and needs, which resulted in awkward political associations between actors that shared neither common political goals nor class background. NGOs proved just as flexible and inventive in connecting with the grassroots—they often wholly invented or adapted programs to fit the requirements of international granting agencies. Li (2015, 85) points out that in the neoliberal economy of Indonesia, national NGOs, like state officials, "cannot maintain themselves, or have an effect in the world, without the funds and legitimation that projects supply." She further considers the project as a device that shapes and aligns the conduct of diverse actors by mediating social and economic processes. While the project produces assemblages, the role that infrastructures and ecology play in this assemblage should also be considered. How does infrastructure influence the form and logic of projects?

Though based on separate data sets and approaching their research objects from different theoretical angles, Li (2005, 2007) and Anna Lowenhaupt Tsing (2005) have written in often very similar ways about the social dynamics of projects that shaped Indonesia during the New Order and after the fall of General Suharto. Both are interested in the social effects of projects carried forward by heterogeneous assemblages of actors. Both look at projects critically and prefer an analysis of their dynamics rather than disputing whether they led to increased freedom or an increased subsumption of human lives under transnational governmentality. Rather, they point to the identities and spaces that are refigured by the trajectories of social movements and development projects. In contrast to state-centered analyses of power, Li (2005, 383) argues that in the context of decentralized Indonesia, governmental "schemes work on and through the practices and desires of their target populations." Power is less visible and gains traction by operating at the lowest level of administration, in the lifeworlds of kampung dwellers. Development schemes are championed not only by state agencies but also an array of other authorities, including NGOs and state-backed organizations. While I agree with these insights (curated from extended ethnographic research in rural Indonesia), I sought to put them to work in urban milieus and see how they would translate. How are "charismatic packages" (Tsing 2005, 228), models for social change that float around the globe, unpacked and put to work in urban settings in Indonesia? And how are infrastructure and ecology folded into these "packages?" The New Order era was full of nationalist "dreams of social change" (216)—wholesome urban development that embraced the kampung. However, *rob* prevention reveals the cracks in the very concept of the kampung as a social unit. Before I illustrate the fragmentary nature of the kampung in Kemijen, I briefly elucidate the many ways in which *rob* itself is figured in the wake of administrative decentralization. In doing so, I demonstrate how divergent epistemologies of *rob* impact the formation of different publics.

Figuring [Out] *Rob*

In the case of *rob,* defining causes and effects is a difficult task. I heard many explanations for the cause of recurrent tidal floods. One taxi driver was convinced that sea level rise was caused by the divine hand to punish human indulgence in homosexual flagrancies, while one of my neighbors thought that ships anchoring in ever greater numbers in the bay of Semarang pushed water onto the shore. Melting ice caps in the North Pole were also invoked as a possible cause. A less spectacular common explanation for *rob* levels is temperature. Some residents I spoke with once speculated that *rob* had failed to occur because of evaporation during a hot spell. The most popular (but less productive) position, however, is blaming the government.[8] Notably, all these causes are invisible.

In this section, I describe the figuring practices of residents whose narrative of *rob* is shaped by locality and levels of visibility of infrastructure. For them, the challenge is to navigate a complex urban ecology and infrastructural landscape and make careful connections between infrastructural failure and biophysical rhythms. At the level of analysis, I question to what extent these connections are informed by narratives of self-help and participatory development. This group of residents is different from Wahyu's activist network, which consolidates dispersed local knowledge of infrastructure in order to influence development projects.

At the level of kampung, flooding in one administrative unit (RT) could be caused by the next downstream neighborhood that just happened to level-up streets earlier in the year. Whatever regional trend in flooding was discernible to residents could be proven insufficiently nuanced by the isolated infrastructural improvement works of certain neighborhoods. If a group of residents had been able to secure a grant for an infrastructural top-up, they might be temporarily relieved of direct interference with *rob*. These fragmentary and heterogeneous investments in infrastructure—raising alleys and thereby changing the direction of the water flow—constantly distorted official representations of *rob*.

Paired with economic inequality and unequally distributed preparedness, predicting the destructiveness of *rob* becomes more than just complicated. The difficulty of knowing the destructive potential of *rob* is reflected in the ridiculous height of new houses and the resulting steep entranceways that owners have to hike up in order to enter their homes. Residents who trust official numbers are hard to find in Kemijen. As there is no scientific consensus on the speed or height of *rob*, the knowledge that some community leaders had of *rob* and water infrastructure placed them in a privileged position vis-à-vis the community and the government.[9]

As *rob* levels fluctuated, the reading of *rob*'s impact on public infrastructure served as testimony to *rob*. While houses and streets may try to keep up with the tide, most public infrastructure doesn't make the cut. Measuring the urgency of repair based on the impact that *rob* has on public infrastructure gives residents not only some clarity as to the speed of sinking; it also gives them political leverage on the district's administration. Old bridges as well as dated riverbanks serve as (slowly disappearing) elevation points against which the changing water levels can be measured (Figure 14). As part of the "public" urban landscape, these elements of water infrastructure often become sites of political debates, which are inflected by specific political sensibilities that shape residential responses to *rob*.

Talut Reading

The *talut* (embankment) represents a recent artifact of (public) water governance that, as I described in chapter 2, was introduced to Kemijen in the late 1980s. Today, concrete walls of varying heights enclose the Banger River all the way from its lowland artificial spring to the littoral. River walls are supposed to prevent flooding by fixing water in place. Conversely, the river walls in Kemijen are seen as a major cause of flooding and variously called out for causing *bocor* (seeping) or *ambrol* (collapsing). Instead of fixing water in place, the banks allow for speculation on their utility. Many residents doubt the consistency of the construction materials or judge the location of riverbank extensions useless. As such,

Figure 14. A riverside resident posted this photograph on Facebook with the following caption: "11 Mei 2015 ... Ketinggian air Rob sudah menyentuh batas bawah ketebalan jembatan" [On 11 May 2015 the height of incoming tidewater already touches the lower extent of the bridge]. Photograph by Joko S. Toto.

riverbanks can become vehicles for political and monetary claims. To illustrate this, I describe the case of the *talut*.

The walls of a *talut* must have holes. Otherwise, the kampung could not be drained of domestic wastewater and runoff during lower tides. Every few meters, a *talut* is interrupted by a water gate that can be manually opened and shut by residents themselves. In Kemijen, these gates are often broken. Yet even when they are in working order, they have become more or less obsolete. At high tide, wastewater cannot exit the neighborhood—the incoming tide lurks outside the gate, so opening it would immediately flood the kampung. Known to be of little use, this older drainage mechanism has already been partially displaced. Many neighborhoods in Kemijen now run small local pumping communities. They close the gates with makeshift mechanisms at dusk and open them in the morning when the tide is low. As soon as the gutter fills up and threatens to spill over, residents resort to diesel pumps to dispose of drainage water into the canal. These communities are an example of a type of local governance regime. They comprise mostly poor residents who have succeeded in soliciting external funds in order to buy pumps. Costs for diesel and maintenance are covered by the *iuran* (communal tax) administrated by the neighborhood head. Those who are unable to pay can contribute in the form of labor (see Kinanti 2013).

Arief's *warung*, where I spent many evenings listening and partaking in neighborhood conversations and gossip, often acted as a space for organizing *pompanisasi*. In the heated period leading up to the parliamentary elections in 2014, Arief's *warung* became a regular site for debates. As I noted earlier, the *warung* was referred to as *dewan*, short for Dewan Perwakilan Rakyat (People's Representative Council)—Indonesia's main legislative assembly. In the confines of the *warung*, customers could have discussions with political content; they could speak openly and share their thoughts about the neighborhood's development and future. As head of RT, Arief is in charge of the local *pompanisasi* pump. One evening, when I inquired about how his business had become such an important local hangout spot, he responded in his usual humble way, by saying that a lot of neighborhood chiefs (RT) and local figures (*tokoh masyarakat*) had started to congregate here at night. In fact, whenever I spent time at his shop, individuals holding some kind of office within the community often stopped by: secretaries, treasurers, and chiefs.

Of course, conversations were not limited to politics and maintenance; sometimes people would show off new gemstone (*batu akik*) acquisitions, gossip about neighbors, or inquire about property prices. While disagreements at the *warung* were encouraged—and in fact promised good business for Arief as customers would often dispute and drink tea until the wee hours—they also brought people together and strengthened their sense of belonging. They spoke freely and felt comfortable expressing critiques of the government and various development schemes. The *dewan* had the advantage of being located at the southern end of Kemijen and forming an important gateway to the neighborhood. It was a lively space that Arief and his constituents kept clean, orderly, and well maintained. They also policed the area, by enforcing traffic restrictions and keeping an eye on strangers entering the kampung. This mimicking of official authority gave them some authority over *rob* assessments and prevention. Their socially and ethnically diverse community practiced an officially sanctioned politics of informal kampung labor that positioned them well in the competition for infrastructural improvement funds. The pump itself represents tangible proof—it was bought with a grant from Indonesia's Program Nasional Pemberdayaan Masyarakat Mandiri, a program I explain in more detail near the end of this chapter.

The coevolution of hydrological systems like the Banger River and street-scale gutter networks show that drainage in Semarang is itself a shifting thing. Infrastructures and communities coevolve in emerging idiosyncratic relations with the environment. While infrastructures and the environment have been shown to be dialectically related, less attention has been paid to the role that local politics play in this process. For example, Anne Rademacher (2011, 16) refers to urban ecologies as the result of "biophysical constraints and social imaginings that converge on a given landscape," and she conceptualizes ecology as a set of experiences and action. Björkman (2014) complicates Rademacher's vision of urban ecology in that she brings attention to the fact that the infrastructures that make up an urban ecology are constantly being contested and manipulated by urbanites across the social spectrum. In this sense, the ways in which inhabitants tweak and fix hydraulic infrastructure produce everyday life in the city. In this process, locality "shores up power and authority" by means of the intimate knowledge that close relationships with infrastructure confer (Björkman 2014, 502). The regular misfortunes of *rob*, as described in the case of the *talut*, have resulted in the emergence of pumping communities that act with a certain level of autonomy from the state. Their engagement with river infrastructure through *pompanisasi* institutions produces a specific local knowledge of water infrastructure.

Local attempts of abating *rob* are not the only thinkable avenue of increasing self-determination and gaining some political leverage in the governance of tidal flooding. The civil organization Komayu, to which I turn now, has strategically positioned itself to be an authoritative voice on the experience of *rob*. Instead of limiting the experience of *rob* to specific localities, Komayu maintains a regional perspective. In contrast to the *dewan*'s role of reproducing locality and neighborliness, Komayu strives to form a coalition across neighborhoods and operates with local and national NGOs. During my fieldwork, the resident-activist Wahyu represented Komayu in the Aliansi Masyarakat Terhadap Rob dan Banjir (Civil Alliance Against Flooding), a Semarang-based initiative presided by the local branch of a national advocacy NGO. These programs attempt to scoop up dispersed *rob*-related knowledge and scale it up.

The *Kampung* in Fragments

Wahyu is a calm and unassuming man in his thirties. Like most residents of Kemijen, Wahyu grew up in the area and maintains a social life structured by close kinship ties. Wahyu joined the ranks of Kemijen's community organizers in 1990. Back then, resident-activists succeeded in "bringing in" many organizations to improve the "dark and violent area" (*daerah hitam dan keras*), as he put it. According to him, "there were deadly knife fights" on his street prior to this time. For many local activists, early involvement with Karang Taruna, the state-backed Association of Active Youth, was a formative experience. The national association focuses on improving the welfare of the poor (Wicitra 2014). In the New Order, local organizations were widely replaced by state-sponsored groups out of the regime's fear of insurgent movements. By concentrating on community development, "these organisations were able to convince the New Order government that they would not engage in grassroots political activities as the banned left-wing organisations had done in the early 1960s" (Hadiwinata 2003, 91). Karang Taruna and other associations were installed with the purpose of absorbing the unruly, criminal energies of local youth. They traditionally operated at the subdistrict level, the kelurahan.[10] After being recruited by Karang Taruna in the 1990s, Wahyu got involved in organizing a subdistrict-sponsored program against domestic violence. Wahyu remembers that he was initially considered "dangerous" and flaky for daring to question the decisions of senior community members. Police officers were sometimes posted right behind him at kelurahan meetings. He thought that people didn't like his criticism because they preferred not to "look for problems" (*cari masalah*). He informed me that in Javanese there is a metaphor for this kind of thinking: *legan golek momongan* (look after your own children). He thought that this mentality still guided residents in the neighborhood, and in his view, this was precisely why *rob* continued unabatingly in Kemijen. Wahyu sees the kampung as shattered into fragments. In his view, it is this divided landscape that makes the governance of *rob* so difficult.

Today, Wahyu organizes community meetings and is regularly received by government authorities. He is a trusted project coordinator; authorities trust him to represent the local grassroots, not radicalize them. Nevertheless, for Wahyu, the problem of the mind-your-own-business mentality remains. In his opinion, the instruments of participatory governance that made communities like Arief's and individual households eligible for subsidies had also helped produce a highly uneven urban landscape in which residents acted solely out of self-interest. Wahyu hoped to close the ranks so that responses to *rob* could become more uniform and inclusive. In view of this splintered network of neighborhoods, Wahyu has tried to develop alternatives to the conventional participatory budgeting and development schemes.

Wahyu lauded the government's legislative efforts to fight flooding, some of which are outlined in the city's master plan (Rencana Tata Ruang Wilayah) and drainage regulation (Peraturan Daerah Drainase). While he thought that they had the potential to successfully curb flooding, he criticized the government for not enforcing the uniform application of such ambitious laws. In particular, while the allocation system Musrenbang (short for Musyawaran Perencanaan Pembangunan) was a "good initiative," in his eyes it also reinforced a disjunctive landscape and society. The national program represents Indonesia's commitment to "decentralized" governance since 1999. Based on a governance brief from the United States Agency for International Development (USAID), UNDP Indonesia (2017, 4) called Musrenbang a "deliberative multi-stakeholder forum that identifies and selects community development priorities." By "negotiating, reconciling, and harmonizing" differences between governmental and nongovernmental stakeholders as well as forging collective consensus on development priorities and the distribution of funds, Musrenbang is supposed to respond to bottom-up impulses for local governance and increase local participation. In line with recommendations from foreign agencies, such as USAID (a major source of funding for many Indonesian NGOs), projects like Musrenbang pioneered "Indonesia's transformation into an open-market economy" (Aspinall 2013, 42).

In Kemijen, a subdistrict with more than thirteen thousand inhabitants whose income levels vary considerably, identifying collective development targets is, to say the least, a significant challenge. Once a year, each kampung can submit a recommendation for necessary "activities" (*kegiatan*). Most community proposals concern Kemijen's roads. Records of allocations of Musrenbang funds confirm this. In 2015 funds were granted for the raising and paving of roughly two kilometers of road infrastructure. Paving (*pavingisasi*) and street raising (*peninggian jalan*) accounted for about 60 percent of the approved funding (INR 277,000,000).[11] Interestingly, the yearbooks that I consulted revealed a staccato rhythm of infrastructural adaptation. A closer look at the records and conversations with administrative staff showed that some of the applications had been pending for at least a year or the allocation of funds had been denied. In fact, competition between communities translated into the tactical placement of proposals. Here, connections to individuals within the bureaucratic apparatus were key for certain neighborhood committees (*kelompok*); ensuring proposals fell into the right hands increased the chances of acquiring funding. Lack of communication between the committees further created disjointed allocative decisions. Had the plans been better coordinated—a rather unlikely scenario—delayed action by the government would still have undermined a collective reaction to rising water levels.

Thus, with good reason Wahyu thought that if everybody occasionally paid a "rent to nature," that is, lifted their streets when money was made available through Musrenbang, they stood no chance at ending regular flooding. He argued therefore to abolish this form of piecemeal adaptation or make the "rent" the same for everybody. Specifically, he argued that subsidizing private construction, or outsourcing government tasks to individual households, created uneven rates of exposure and vulnerability. Wahyu added that the subdistrict was further divided by political, class, and religious differences. He blamed the manifestation of these divisions on the politics of decentralization, which had underscored, if not unleashed, the selfishness of residents. Moreover, Musrenbang was only focused on top-ups (peninggian) of existing infrastructure. His own organizing strove to undo the spatial fragmentation of flood prevention by modifying the role that grassroots organizing played. In short, his goal is to achieve a uniform and timelier adaptation to flooding trends that would give all residents the same chances at maintaining their homes while also reducing *rob* more generally.

Rob and the Grassroots Structure

In Kemijen, rumor had it that the government would soon begin evicting residents by the fishpond, as it was slated to become a water retention basin.[12] Wahyu was unhappy with the situation. While his neighborhood would stay unaffected by the evictions and potentially benefit from reduced flooding should the basin be implemented, he took issue with the "nontransparent" procedure of government agencies. At the same time, he was also critical of the fact that residents hired a private lawyer (*pengacara*) in an effort to contest the infrastructural project. As a result, the lawyer was dealing with the government on their behalf—as intermediary. Wahyu thought that they should represent themselves as a united group of residents. In his view, they would then be able to have a stronger voice and insist on a transparent and fair procedure. Involving a lawyer was dangerous since he could make deals with the government behind closed doors. Who knew? If the "guy" accepted legal fees from residents, he might be persuaded to accept payment from the municipal (*pemerintah*) or local government (kelurahan) as well. Wahyu did not believe in such messy negotiations—it was the "old way" of doing things, where every person involved in the deal wanted a piece of the project money. There were too many unknowns and temptations to cheat in this process. Wahyu didn't trust the authorities altogether, but he did trust certain individuals within their ranks. In his spree of activities, he was often forced to deal with existing structures of local governance and to be on good terms with the authorities.

I knew Wahyu as a terribly busy man. Specifically, if a program was about infrastructure, economic development, or flooding, he played some part in it. He was often directly responsible for the dissemination (*sosialisasi*) of information concerning a new kampung-scale program to residents; he also always helped organize and coordinate these programs. His voracious organizing could at times give way to visible fatigue and silent resignation. Having to constantly probe and push the boundaries of what was feasible in the arena of kampung development, his moods could shift on a whim. Wahyu focused on local projects; I rarely saw him working on projects whose scope surpassed the subdistrict level. In a tangible sense, the grassroots structure both enabled and constrained his everyday work. One day, he could be vibrant and full of hope about new local development opportunities. The next day, he could be found shaking his head in despair, rubbing his sleepy eyes as he complained to me about the "endless struggle" and the "uselessness of it all."

(Closeted) Activists

In 2000 the activist group Komayu was formed to address two major issues of local concern, flooding and economic development. Today, the group has a chief coordinator and consists of a variable number of informally nominated members who can be elected to special offices. Since its formal foundation, Komayu has cooperated with several national and international nongovernmental organizations. Komayu's operating budget mainly derives from project budgets of donor-based organizations, whose grants are often made dependent on cooperation with the urban grassroots. Komayu has managed to attract funds from several sources, gaining influence for its members to participate in the decisions of the subdistrict government about how and where infrastructural programs get implemented. Wahyu was recently elected chief coordinator of Komayu.

Previously, Komayu was not regarded as a legitimate partner in local development. It had to craft this image over time. As an experienced and full-time local activist and labor union member, Adin (introduced in chapter 1 and portrayed in more detail in chapter 5) helped organize a "folkloric" festival from 2006 to 2009. These events aimed at putting the organization "on the radar" of society and the government, and to therefore boost its ability to partner with government agencies. Its members wanted the government to see that the communities along Banger River were able to represent themselves and become partners in bottom-up urban development projects. Wahyu said that the municipal government took notice: "Oh, look at that, even the stinky Banger River can do it" (*bahkan Banger yang bau bisa*). According to Adin, attracting development programs was the fruit of their continued efforts.

Wahyu explained that the reason for creating Komayu was a shared desire to change Kemijen. They began by celebrating an annual *ruwatan* (a traditional Javanese cleansing ritual) by the river to bring about positive change. Adin pointed out that the event was primarily aimed at attracting attention from government officials. According to him, the ceremonial aspect represented the community's "readiness" for improvement to the wider public and a distinct cultural "package" (*digemas budaya*) to appeal to developmental interests. In fact, the group's members learned to put extreme care into how their message to the public was "packaged." Their critique always had to be subtle, which required the early involvement of the government in their activities instead of other more openly "political" means, like protesting or demonstrating. They feared that the government would blame them for troublemaking and withdraw support. In the past, the government had legitimated the absence of development projects by pointing to political "elements" (*elemen*) in Kemijen. Involving the government was thus a preemptive act that eliminated or reduced suspicion. Group members recalled that a few years prior, they became the object of a police investigation due to organizing "dark meetings" (*rapat gelap*). When the group sensed the specter of criminalization, Komayu started operating through more dispersed networks. For example, Komayu now cooperates with a local advocacy organization whose activities are financed with international donor money. Wahyu informed me that "you need a lot of friends if you want to take on the government. . . . Your arguments need a foundation." By demonstrating visibility and staying connected across the political spectrum, the group seems to avoid being criminalized for voicing critique.

Grassroots groups need to be loosely connected with political camps to avoid suspicions of corruption or being categorized as too militant. These groups also have to signal adherence to traditional local hierarchies. In fact, Komayu has managed to insert its members into numerous legitimate sites of local budget dispersion. At a routine Komayu meeting that I attended, Wahyu was surrounded by his many allies. He looked confidently into the roundtable of assembled members: "Mas Toto is head of PNPM [a government-initiated poverty alleviation program]; Pak Tanto is head of KSB [disaster-preparedness group] and RW (neighborhood association)." The list continued. After Wahyu's remarks, Adin smiled and added that their work and infiltration into donor bodies resembled a constant "gnawing" on the government, "like mice!"

The global discourse of climate change served Komayu as a vehicle for pressing for increased residential involvement in urban planning. In 2010 Komayu organized a climate change festival as part of an "awareness raising" campaign. Adin directed workshops on climate change and its potential impacts on livelihoods to civil groups and residents. He knew that the area's vulnerability to climate change had been established by government-backed vulnerability assessments carried out by the international NGO Mercy Corps, whose study predicts that about 185 hectares of Kemijen's territory will be permanently flooded due to sea level rise (ISET 2011). Adin recalled that the festival produced tangible results, including the annual renovation of *tanggul* (river walls) by the government. However, explanations of *rob* based on climatic trends did not stand the test of time in Semarang.[13] As further examples of Komayu's work will show, it was precisely the impossibility of pinpointing the exact origin of *rob* that made engaging the government possible. Climate change, as a universal claim, thus served to promote specific local demands (see Choy 2011), such as punctual infrastructural repair.

Treading Carefully: From Festivals to Development Plans

Organizing festivals helped maneuver Komayu into a strategic position to communicate with sections of the government. From here, the group has been able to exploit people's uncertainty about the causes of flooding to increase its political legitimacy. This in turn allowed the group to become one of the accepted voices of the area's residents in municipal participation programs. Wahyu was hopeful that partnering with national NGOs would help marshal a number of bottom-up initiatives and gradually make their voice heard at higher levels of the government. With the support of the umbrella organization Pattiro, a *jaringan* (network) was formed that consists of members of Semarang's Lembaga Bantuan Hukum (Center for Legal Assistance) branch, journalists, academics, and residents of Kemijen. Under the name Civil Alliance Against Flooding (Aliansi), this network has called for a series of discussions with municipal stakeholders, to which I briefly turn in order to present an example of how Komayu and other activists stage Kemijen as an endangered area populated by well-intended grassroots actors ready and willing to help spur economic development potential.

In 2014 Aliansi invited municipal stakeholders and civil groups to discuss the newest drainage regulation (Peraturan Daerah Drainase) and the ongoing antiflooding measures being carried out by three municipal government bodies in a "social audit." Semarang's water agency (PSDA, Pengelolaan Sumber Daya Air), environmental agency (BLH, Badan Linkungan Hidup), and planning agency (Bappeda) attended the roundtable discussion. Impressively, the network had managed to assemble *rob* victims, activists, and decision makers in the same room. At the meeting, budgeting and transparency in infrastructure provision were salient topics. The chairperson stated at the beginning of the meeting that the "conversation" was going to be strictly about empirical "facts" and not feelings. The aim was to give the government a kind of reality check. As it turned out, actual facts were difficult to produce. In one instance, a participant complained about unfinished roadworks in his neighborhood where the contractor (*pemborong*) had done a poor job. The participant said that he contacted the council member who had ordered the renovation, but the man only offered to meet him in his office. "I wanted to show him [the council member] the facts on-site [*di lapangan*]—how else could I prove the facts?" the resident asked, barely concealing his discontent. Only at a few points did the activists manage to present evidence of minor miscalculations and fraud. Why had renovations of this stretch of road not been completed, as described in the publicly accessible construction contract? Why did another contract dubiously stipulate the need for an ominous insurance policy? However, these pushes for rectification of fraud were not accompanied by outrage, and no offense was taken on the government's part. For example, a young man spoke about a riverbank that PSDA had recently put in place. Apparently, a stingy operator had used way too much sand in the concrete mixture. In fact, the complainant had brought the evidence with him. Spectacularly, he dropped the entire contents of a plastic bag on the meeting table. Small pieces of concrete burst out on the table and broke into even smaller pieces. The man went on to demonstrate the extreme porosity of the material by pulverizing it between his fingers and throwing the remainder in the air in a dramatic gesture. The gesture seemed to scream: how is this riverbank supposed to protect against flooding? But instead of eliciting feelings of anger due to a sloppy construction job, everybody broke out in laughter. In fact, the officials had no choice but to indulge the well-staged comedy of state incompetence, probably suppressing more defensive reactions. After three hours of "auditing," which contrasted government plans with actual outcomes, Adin said in a reconciliatory gesture to the representative of PSDA: "Please don't think that we residents are against the government. We want to work together." At the meeting, both sides were keen on keeping their collaboration alive at all costs.

A representative of the chairing NGO Pattiro told me that Aliansi was supposed to become a "critical partner" of the government. To do so, the NGO had to collaborate with local residents in order to identify and relay their "actual needs." As critical partner, they would be able to bring society's needs (*membawa kepentingan masyarakat*) to the attention of the government. The government should "get the facts from civil society" (*fakta dari masyarakat*), the representative added. What became prevalent in their mediation of facts was the specific role that kampung residents had to play. Achieving "maximum effect," that is, influencing the discourse around *rob*, would not come from a conversation between the government and the NGO. Aliansi's legal expert reminded them at a meeting that "flooding" (*rob dan banjir*) was in any case already a "reality issue" (*sudah isu realitas*). Though undoubtedly "real," in his view they rarely had a say in any plans that addressed the issue. From an interview with another member, Adithia, a local journalist, it became clear to me that how one spoke about facts was more important than simply producing them. *Rob* was a "fact that could not be hidden" (*tidak bisa disembunyikan*). Aliansi therefore had to present a specific story of *rob*. Kemijen was a "strategic" space from which this story could be launched into wider political discourse, as all members agreed. Presenting the area as undergoing substantial cultural and environmental change via spatial and infrastructural improvements was successfully attracting ever more governmental and media attention (see chapter 5 for a discussion of the art collective Hysteria's project). There was disagreement about the content of that story, however. The fact sheet that Pattiro ended up presenting to the government based on its audits was strongly criticized and rejected by municipal staff. Particularly, the language used by the NGO was judged inappropriate and unacceptable. Wahyu told me that this had happened before; in his view, the young NGO often didn't know the proper approach to dealing with the government.

Having an influence on water and infrastructural governance in Kemijen is a delicate process in and of itself. Claims have to be packaged and dressed in the cloak of grassroots culture. By turning their precarious surroundings into a culturally unique "endangered environment" in contradistinction to dominant visions of the neighborhood and area (see chapter 2), Komayu's members have been able to speak to delicate social issues, such as equal accessibility to reliable drainage infrastructure and political representation. While climate change festivals aimed at raising awareness of environmental issues evoke the image of the uninformed downstream kampung inhabitant, these events coincide with the desires of democratic self-determination expressed in numerous infrastructure and development projects. Collaborating with the NGO Pattiro provided a way out of social and political isolation. Yet the "grassroots structure" can also put the government and overly ambitious NGOs at loggerheads. Representative of Komayu's larger efforts, Wahyu had to tread carefully so as to not reanimate the specter of civil disobedience.

Pembangunan versus *Peninggian*

Wahyu always tried hard to stay in the loop of phone calls, project cycles, and meetings. I could sympathize with his occasional overwhelmedness as I struggled to remain cognizant of all the various projects underway at that time as well. In addition to participating in a series of audits with the municipality organized by Pattiro, he was the local coordinator of a kampung development program. He also coordinated infrastructural assistance claims made by his neighborhood association (RT), helped promote a bottom-up drainage maintenance program, and organized public events to improve Kemijen's image. The following observations on his involvement in community-based infrastructure project speak to Wahyu's newly gained position as local representative of the neighborhood. While I was conducting fieldwork, he was particularly invested in one infrastructure program because it seemed to afford a different kind of adaptation to *rob*.

In 2014 Wahyu was elected coordinator of a community-based infrastructure project that was financed by PNPM—"the mainstay of Indonesia's poverty reduction program" and ground zero of participatory planning initiatives aimed at reaching the United Nations' global Millennium Development Goals (Li 2015). Wahyu was charged with overseeing and advancing the design of an infrastructural intervention that would "regulate" (*tata*) several neighborhoods. According to a newspaper article covering the program, the subdistrict was so "filthy" (*kumuh*) that it "really needed to be regulated" (*Suara Merdeka* 2015).

A small urban park was supposed to replace an eyesore of the local landscape: an unregulated garbage dump between two kampungs located east of the Banger River. Wahyu's team hired a planning professor from Semarang's biggest university, Universitas Diponegoro, for the job. Upon seeing the design, I found the plan overly glossy and almost clinical to the point where the street depicted in the design bore no resemblance to the actual place. I asked his assistant why they preferred square concrete blocks to planter boxes that mimicked the riverbank's (*talut*) streamlined structure? He answered that they foremost needed a cleaner (*lebih bersih*) and more orderly neighborhood. Improvement of the neighborhood's overall appearance was a condition for further development. The unrulier a neighborhood looked, the less public investment it could expect. I immediately felt bad for being such an urban design snob. In fact, the grant timeline didn't allow for much creativity in design; the program would be audited shortly and they had to spend all the funds by a certain deadline. After a few months of planning, Wahyu came under considerable pressure to deliver a feasible design backed by locals, one that would please both municipal planners and the organization providing the funds—the latter wanted to be sure that the program was implemented in a truly participatory way. As a result, after work Wahyu often drove around on his motorbike to distribute formal invitations to neighborhood heads. Unfortunately, the meetings had increasingly poor turnouts, despite his team showcasing well-crafted PowerPoint presentations from the consultant at each meeting. Since turnout stayed minimal, Wahyu's team used every opportunity to generate interest and public approval for the project. For Wahyu, it wasn't just fulfilling the program's conditions for minimum participation that drove him to reach out to his neighbors. Wahyu was hoping to generate real shared interest in the project: interest that could build a common vision for the fight against *rob*.

Wahyu hoped that a Festival Rakyat (People's Festival) would ultimately put people's support behind the project. It was common practice to organize festivals because they attracted many kampung inhabitants. This festival took place in the proposed project zone, on the east side of the Banger. On the day of the festival, between deafening *dangdut* performances (a popular type of music), lotteries, and speeches, organizers announced the project. Somehow Wahyu found time to commission a giant banner for the podium that showcased images of the planned infrastructural changes. The plan, which bore little or no resemblance to the present condition of the area, foresaw the construction of a badminton field and garden (*taman*) as well as a riverside promenade with sleek sitting benches. Providing the infrastructural elements of an "urban park" was supposed to improve sanitation in the area, which, according to a preconstruction survey, consisted mostly of "semipermanent" and "inappropriate" (*tidak layak huni*) housing (BKM Kemijen Mandiri 2014).

During the festival, I climbed the stage and sang a Western ballad accompanied by my accordion-playing friend. The enjoyment was great when a transgender *dangdut* artist asked me out on a date in front of the one hundred festival attendants. Then, the lurah held a speech, and his wife partook in a partly improvised performance of *wayang beling* (a traditional Javanese stage play in which actors speak in low Javanese and are allowed if not expected to swear). The new *camat,* the head of East Semarang District (Kecamatan Semarang Timur), swung by but successfully avoided a speech. He admitted in a brief conversation that he had shown up without knowing anything about what the festival was trying to achieve. Like most other attendants, he was there to celebrate the bestowal of an infrastructure grant for the area. Events like these were well-rehearsed encounters between residents and the higher echelons of the local administration. They meant free entertainment for children and adults and produced welcome income for local performers and sound engineers. However, the masses' amusement was neither understood as a sign of standing together or creating a sense of community, nor did it improve residents' knowledge of the project; it merely served to make the decisions made by the leadership more official. Wahyu was disappointed by the meager turnout. How were they going to form a common vision if residents showed no interest in planning?

However, the meeting dedicated to planning the festival revealed that Wahyu had bigger fish to fry. From the beginning, Wahyu wanted to set a different tone for this event (I briefly describe the planning for this festival in chapter 2). As a tentative title, he proposed Festival Kepedulian Kali Banger to the event committee. He explained that they had to add a human touch to the festival. The previous title, Festival Kali Banger, didn't capture the new spirit of Kemijen: the festival had to signal an awakening of the local population, a readiness to come together to care for the environment. Unfortunately, the meeting was ultimately reduced to deciding who would appear in the comic theater performance. To the lurah, it didn't matter what they called the festival as long as it attracted people. He promised that if people knew the program well, they would probably support it (*kalau masyarakat sudah tahu persis pasti dukung*). As I showed earlier in this section, "knowing" the program didn't refer to a detailed technical understanding of the infrastructural additions to the subdistrict. Rather, for the purposes of the event, "knowing" was limited to acknowledgment of the project.

Conversely, Wahyu looked to imbue infrastructure with a new meaning—to make it an object that people cared for communally. I suggest that Wahyu was retrofitting an ideology that was key in Indonesian developmental nationalism, whose idea of technological progress "was located within the project of nation building" (Barker 2005, 710). As Barker (2005, 710) has pointed out, the term *development* can be translated into either *perkembangan* or *pembangunan* in Indonesian: "While *perkembangan* implies a development associated with flowering and expansion, *pembangunan* suggests that development is a process of awakening, building, and construction." Wahyu's festival title conveyed this vision of an awakened community. Wahyu implicitly contrasts a conception of development through "construction" that stems from public awareness with one obsessed with material infrastructure. Ariel Heryanto and Nancy Lutz (1988, 22) have shown that pembangunan not only "works at renovating the life of society, but also opposes, rejects, and eradicates anything which is considered 'traditional' or 'anti-modernization.'" To do so, pembangunan entails a recruitment process through which members of the society are rallied behind a future-oriented project. Wahyu is nostalgic for development that elicits change at the societal level and gets rid of the "old way of doing things." This vision of development is diametrically opposed to the current way of leveling up kampung roads, suggested by the term *peninggian*.

Epilogue: A Sensible Politics of Water Management— and Its Outcomes

The local joke—"paying rent to nature"—is a poetic and stoic way of describing and normalizing years of living in a precarious relationship with a rapidly changing environment. This relationship is framed in monetary terms by residents, which gives it the exclusive contours of a contract. Moreover, it signals a relation of trust between inhabitants and nature; inhabitants want to actively protect and nurture this relation. But this form of ecological membership is increasingly made exclusive through disparities in the local distribution of water and access to water infrastructure. Not all residents can take an active part in defining their relationships with water. Advancing democratization and decentralization in Semarang has given birth to new approaches to local development—civil organizations, public–private partnerships, and advisory boards. While they encourage civil participation, they have not undone local dynamics of uneven development; rather, they have played into them (Li 2007).

The circumstances of Ariel and Arief as well as Wahyu are well worth revisiting here. Ariel grew up in Kemijen and recently inherited land from her parents. The vacant plot lies less than ten meters north of Ariel's current home, which is regularly flooded due to its low elevation. Her husband and her envision building a house on the higher plot; perhaps they will even give up their current home as lost to an accelerating ecological trend. Through their ability to attract grants as responsible kampung members, they are able to envision a different future. While they are undoubtedly also building on borrowed time, this kind of borrowing functions in a different way, through appropriation. They appropriate the time afforded by state funding in order to carve out new opportunities. In Kemijen, time inheres in the different programs and project structures made available to residents via a structured political hierarchy.

Through paying rent to nature, a new affective attachment to space and nature can be signaled. This chapter has shown that certain communal water management projects are intimately tied to long-standing ideals of democratic reform and can even prove successful in undergirding residential rights to urban infrastructure (Bunnell et al. 2013). A local structure of feeling is linked to a growing attention to infrastructure, a symbol of both promise and failure through which ambitions of communal governance are enacted. The case of Kemijen demonstrates that local political and ecological practices reflect an engagement with the environment that puts Semarang's northern kampungs in a new light. In essence, water infrastructure is provided if it is managed by culturally oriented and responsible citizens. The emergence of a local water politics in Kemijen illustrates the political contours of an ecological disaster, *rob*, in an empirically specific way. We need to consider responses to environmental challenges as both fixes of political disorder inherited from the past and imagined solutions to unsolved problems of social membership. While projects mostly shore up the precarious existence of residents, they can also generate beneficial collaborations between unlikely partners.

Wahyu's efforts to seek out unlikely partners exemplifies how the grassroots structure found new articulations through *rob*. Wahyu, backed by Komayu and its citywide network, envisions an alternative to recursive *peninggian* development. While Komayu's organizing does help certain residents find a new, and overall more prominent, position with regard to local development, it is questionable whether it allows them to have real, democratic impact on the kind of development help that flows into Kemijen. Within the grassroots structure, there are more powerful actors, such as members of the municipal administration, and major development agencies, such as the World Bank, who manage development schemes, such as Musrenbang, and so inevitably reproduce "project time."

5

PROMISE
REMODELING DRAINAGE

In 2015 the Semarang-based art collective Hysteria organized Penta K Labs, a water-themed art festival, in the streets of Kemijen.¹ The collective had been strategic about the selection of the neighborhood. Kemijen was notoriously afflicted by *rob,* and it was common knowledge that the local government was collaborating with Dutch experts in the hopes of getting *rob* under control. Hysteria's intervention sought to draw attention to water management problems and display residents' ingenious efforts to cohabit with flooding. The director of the art collective insisted that local residents knew exactly what they wanted the government and media to take away from the festival. They sought to counter the public opinion that northern kampung residents were apathetic and dangerous and were in fact more than ready to cooperate with the government in antiflooding programs. Figure 15 depicts an installation piece that created a riverside hangout out of a dysfunctional public pump. The mural reads: "Keep strategizing until the polder becomes reality." The event was a double success. Hysteria reaped critical acclaim in local media (Mingu 2016) and increased its overall visibility, while Penta K Labs successfully assembled international artists and acquired local support with Semarang's incumbent mayor, Hendrar Prihadi, in attendance. As a result, foreign organizations who run resiliency programs in Semarang, such as 100 Resilient Cities and the Rijksdienst voor Ondernemend Nederland (RVO, Netherlands Enterprise Agency), have since partnered with the collective. Moreover, the attendance of the mayor was seen as a positive sign by Kemijen's residents. More than three years later, Wahyu proudly remembered how the event gave him his first chance to speak with the mayor about issues that the government urgently needed to address.

The polder is a Dutch–Indonesian drainage project aimed at ending tidal flooding in neighborhoods flanking the Banger River. Dutch water engineers and policy makers, the central government, and Semarang's municipality worked together to stop *rob* in North Semarang and reconfigure local water governance at the same time. Together, they promised "dry feet" for approximately eighty thousand residents. Notably, while the project was masterminded by hydrological experts in Jakarta, the actors involved in its implementation came from various socioeconomic backgrounds, and many key actors had not studied water management and/or lacked experience in urban governance. This "multi-stakeholder" approach was unprecedented and suggested a democratic

Figure 15. Street art along the Banger River in North Kemijen. Artists turned a small defunct pumping house into a hangout spot. They wrote: "Keep strategizing until the polder becomes reality." Photograph by Lauren Yapp, 2016.

opening of urban water governance in the face of Semarang's crumbling drainage infrastructure. Despite professional, cultural, and social differences, project partners unanimously agreed that the city's drainage system was unable to cope with present ecological conditions. At least initially, they envisioned a rupture with the usual approach to infrastructural upgrading—the process of peninggian, which consists of regularly stacking up streets and riverbanks to outpace the rising tide. The polder project was supposed to help communities directly affected by *rob,* whose struggles with poverty were compounded by recurrent floods. The aim was to undo the messy temporality of informal flood prevention in the form of riverbank fixes, pumping communities, and house repairs. Most residents enthusiastically greeted the intervention, as it promised to remove the stain of backwardness and economic stagnation from the area.

This chapter examines the polder project from a critical perspective. What happens when a consortium of Dutch and Indonesian experts tries to remodel the drainage infrastructure of a vast inhabited urban area? In particular, I reflect on the ways in which framing the situation as an ecological crisis allowed for unprecedented levels of cooperation between the governments and residents. Taking stock of the project at different moments of its coming into being—2014, 2015, and 2018—I show that the notion of a flood crisis provided a rare opportunity for increased residential participation in deciding the fate of Semarang's floodplain. This crisis scenario led to the installation of a unique water authority, SIMA, which allowed certain residents to be involved in the implementation of the polder system.[2] In particular, by establishing SIMA, residents from affected neighborhoods and local academics were able to become stewards of a new type of flood governance. Together, they continue to dream of a different kind of flood management: one in which residents play a major role as both stakeholders and stewards.

As I will illustrate, the project failed to live up to the promise of delivering a "bottom-up" drainage system. Ultimately, it could not shake the normative politics of infrastructure projects and a historically calcified style of water governance that depicts poor people as both incapable of assessing risk and developing a technical understanding of drainage. As such, the project ended up reproducing a cultural imaginary that segregates the North and is itself expressed in the spatial hierarchy of Semarang. Sadly, after the Dutch project partners understood that their vision of a resident-owned polder had failed to take root in Semarang, they began to search for new project partners within and outside Indonesia, leaving it to the politically weakened institution SIMA to fend for its ideal of participatory flood management. While the project leaders had promised real change and increased levels of resident participation, the polder project failed to bring about democratic flood control and participatory urban development. The intended beneficiaries of the project, poor residents who have endured tidal flooding for decades, will eventually leave Kemijen as land prices rise. When they do, they will leave the neighborhood with the impression that co-ownership of water infrastructure is impossible and that even the involvement of foreign engineers doesn't change how the government treats the poor. In Semarang, as elsewhere, a dried floodplain provides land only for those who can afford it. In view of this, I ask more broadly what we should make of international development projects whose legitimacy in managing risk relies on the depiction of human and nonhuman life worlds in crisis.

As James Ferguson (1999, 251) has argued, the forms of political and economic organization captured in the term *development* are often laid on top of "already-existing geopolitical hierarchies." In this sense, "development" neither creates social inequalities nor undoes them; rather, it helps manage and sustain them. This explains why, according to Ferguson, "dead ends of the past keep coming back, just as the 'main lines' that are supposed to lead to the future continually seem to disappoint" (251). The polder project was imagined as the beginning of a new era of urban water management in Indonesia. Instead it repeatedly felt like another dead end, a futile stab at changing the logics of capitalist urbanization. In the following, I reflect on specific signs of the polder's early social and material ruination, revealing each time the effects and limits of crisis narratives as well as the frustration that these "dead ends" caused to different people working on the project. Although a sense of urgency and commitment reigned among residents and project partners alike, the material and social stagnation of the polder raised doubts as to the feasibility of a Dutch-inspired polder in Semarang's North. Furthermore, incredulity and frustration came to supplant a sense of hope as the government lost patience with the polder's participatory design. The project's repeated delays eventually led to attempts of one project partner,

Semarang's water agency PSDA, to co-opt the project, indeterminately suspending its core communitarian approach.

The slow dismantlement of the project's ideological hopes demonstrates the power of the chronic—as a once ambitious development project that promised social change was sidetracked by the status quo of river normalization. This "main line" of urban development, imagined as infrastructural cure, continues to disappoint both residents and supporters of a more democratic city. In keeping with normalization as the dominant framework of urban development, the government undermined the innovative social design of the polder project and reaffirmed a top-down approach to water management. As such, it had merely used the language of localized, participatory development to push an old infrastructural agenda.

Semarang's Polder

In technical terms, a polder is a hydraulic infrastructure employed to manage flooding; it is a hydrologically closed system consisting of dikes, dams, and water pumps. Typically, a polder is meant to prevent tidal intrusion and make living below sea level possible by damming a river or canal where it stretches into the sea. In this sense, it is an expert technology planned and implemented by trained specialists. As a contemporary product of the legacy of Dutch colonialism, many Indonesian cities have implemented polder systems, which are operated by state agencies and often financed—including design and construction—with loans from foreign investors or development agencies. Semarang's municipality has built a total of five polders in the past decade, the first being finished in 2010 (Kementerian Pekerjaan Umum dan Perumahan Rakyat 2019).

The polder system originated in the Netherlands, where the long-held and widespread practice of land reclamation now makes them a common feature of urban landscapes. A polder is surrounded by dikes or embankments that allow undesired water to be controlled either by being run through a sluice or by being pumped out. While polder technology "is a triumph of technology," as M. R. Brett-Crowther (1983, 157) notes appreciatively, creating polders has always been a political undertaking as well. Polders have been described as politically self-contained systems in which the interdependency of all constitutive elements (land use, demographics, economy, etc.) is acknowledged and regulated. Roy Kraft van Ermel, a Dutch retiree with Indonesian roots who visited Semarang repeatedly as a diplomatic envoy of the Dutch side, often said that the Dutch commodity in demand was not simply water infrastructural technology but the know-how required to operate and, more importantly, maintain it. Kraft van Ermel, who had a career in municipal politics in the Netherlands, argued that building polders required thinking holistically and comprehensively while remaining focused on the long term.[3] He surmised that this thinking accrued from living with flood risk, like a hydrosocial apprenticeship, constituting a kind of situated knowledge. An appropriate design of a polder for Semarang therefore needed to pay heed to the interrelation of economic and social sectors, private and municipal interests, and geological and climatic conditions.

In March 2009, the design of Indonesia's first democratically steered polder system was cheerfully celebrated at Novotel, one of the city's most prestigious hotels, located on Jalan Pemuda, the Champs-Élysées of Semarang. All the feasibility studies as well as a "Detailed Engineering Plan" had been accomplished at the beginning of the year. The Dutch water engineer Herman Mondeel spent three years in Semarang devising a technical plan that was both feasible and affordable, as the Indonesian government had to pay for the infrastructural components of the polder. Mondeel, who brought his wife and children to Semarang, learned Indonesian, and became good friends with the members of the then provisional polder board, recounted to me how the available budget had put considerable constraints on his imagination. His employer, the Dutch consultant company Witteveen+Bos, who employed Mondeel and is known for the titanic undertaking of reclaiming land in the bay of Jakarta, won the tender to create the technical design of the polder. The company made the Dutch water authority Hoogheemraadschap van Schieland en de Krimpenerwaard (HHSK, Regional Water Authority Schieland and the Krimpenerweaard) an unbeatable deal, even outmaneuvering Indonesian competitors.[4]

According to Mondeel's design, the polder infrastructure was to consist of a low-budget ensemble of dams and dikes, a pumping station including five pumps, and a water retention basin. On the one hand, sediment dredging and transferring excess water to the nearby flood canal was supposed to lower the water level of the Banger River by approximately 1.7 meters (Witteveen+Bos 2014), restoring Banger's historical water volume. On the other hand, a dam positioned where the stream arrives in the estuary and a dike along the northern edge of Kemijen promised to stop *rob* by blocking out sea water. The Banger River would be reborn as Banger Polder, a hydrologically closed drainage system. The rationale of the design could be interpreted as follows: instead of allowing the floodplain infrastructure to continue to be subject to multiple fragmentary interventions from both state and local populations, the Banger Polder would synergize energies, efforts, and investments. It would also put a single governing entity in charge of monitoring water flows within the polder. The new system would thus mark a shift from endless on-demand infrastructure repair to sustainable water management and an end to seasonal and tidal flooding. Mondeel claimed that the technical design strove to create an infrastructure that could nestle itself seamlessly into the social geography of North Semarang. However, the technical components of the polder—dam, embankments, and pumps—significantly shaped social arrangements and boundaries, especially in Kemijen. For instance, the installation of the largest components of the polder, the pumping house and a retention basin, required evicting families and changing land uses.

The design was predicated upon two basic understandings of the flooding problem. First, it argued that land subsidence had rendered the existing drainage system obsolete. Second, it predicted that climate change would worsen the situation through local sea level rise and increased monsoon rainfall (Witteveen+Bos 2014). The engineering challenge was therefore twofold: blocking tidal inflow while retaining surplus water before pumping it out. The design identified two main external forces impeding "normal" water flow and proposed a primarily hydraulic solution to the flooding problem. Accordingly, it envisioned two interrelated interventions into the urban fabric of the northernmost neighborhood, Kemijen. The first intervention concerned water flow. Pumps were needed to unmake the informal hydrological relationships of riverside kampungs with the Banger River. Put crudely, with a fleet of five pumps operating, neither strong rainfall nor drainage water should be a concern to the floodplain communities, as the water flow would be constantly monitored and hydraulically enabled. In fact, the polder's centerpiece, a pumping house, not only promised a leveled flow of water but would also control groundwater levels.

Maintaining low water levels was supposed to provide subterranean stability. The planned polder therefore not only promised to stop flooding but also set out to create the right conditions for unbridled development activities in the area. Interestingly, the designers admitted that lowering the groundwater level would lead to increased land subsidence, thus accelerating already problematic soil settlement.[5] The design, however, deemed the rate of settlement postinstallation "acceptable, considering that the present land subsidence is 9 cm/year" (Witteveen+Bos 2014). The way in which the pumps were envisioned to perform claimed a detailed understanding of local ecological processes and painted a future in which land subsidence was still a problem, but one that could be dealt with technically, provided that both state and local populations continued to invest in the land.

As the project leader, Johan Helmer, explained it to me, the polder infrastructure was, in any case, only "buying the government time." A mere infrastructural upgrade as envisioned by the polder project would not do the trick. More significantly, the intention of the polder project was to disrupt the current method of flood prevention and introduce a new set of relationships between the government, kampung residents, and water. It hoped to undo specific urban rhythms or temporal configurations by offering a window for institutional changes. Both the Dutch and the Indonesian engineers knew that relief from flooding was going to be temporary. The delta was in rapid decline, and global warming would cause a significant rise of the sea level. Therefore, the polder design was to carry forward a new form of flood governance—one that was different by virtue of its participatory design. Not technology (pumps and dams) per se but specific planning technology, local stewardship, and entrepreneurial management would help prevent disastrous flooding in the future. In a sense, this new relationship with water would also be built on borrowed time—the precious time window afforded by the social and technical intervention of Dutch experts.

As I will demonstrate in this chapter, after the polder design had been released and the location of its infrastructural components revealed, residents didn't immediately stop their own flood-prevention acts. Some even built houses in the territory slotted for construction, in hopes of receiving compensation when the government moved to evict them. They used the polder plan as a "vehicle of action" (Baxstrom 2012, 138). Paradoxically, the polder design and its implementation, which were themselves supposed to buy time for a deliberate shift in water management, also resulted in noninstitutional strategies of borrowing time that undermined the polder project.

Pilot Democracy

Pilot projects are often framed as experiments in that they aim at pioneering a new approach to long-standing problems. The Banger Polder was supposed to test a different style of flood prevention, one that involved the local population in maintaining and operating water infrastructure. In fact, the project differed strongly from other government schemes in that it openly aimed at devolving substantial authority over operation and maintenance to a local body with civil representation (two board members were residents of the project area). As hinted at previously, the various institutions behind Banger Polder wished to end the flood crisis not by using expensive technology and radically reshaping the urban fabric but by involving local residents in planning and running a rather modest water redistribution system. As part of their participation, residents of contiguous subdistricts were expected to pay a water tax after implementation, modeled on the Dutch water levy, to allow for a financially autonomous "polder board."[6] As a legitimate governing entity with spatially limited powers, this new authority was supposed to become a state-sanctioned trustee responsible for levying the water tax and overseeing the operation and maintenance of the polder.

In 2014, when I started my research, a provisional polder authority called SIMA was already working in an office located in Semarang's city hall. As a potentially unique and, for Indonesia, historically unprecedented body of governance, the board was attracting national and international media and some scholarly attention (see, e.g., Richter 2020). This semigovernmental polder authority was an implicit critique of the state for having failed to offer timely responses to the flood crisis. It was conceptualized as an urgent response to flooding that strove to eliminate the contradictions of entrenched practices of water management by pairing Dutch planning technology with standard drainage infrastructure.

Specifically, by creating new spaces in which residents were supposed to inform themselves about the polder and their responsibilities as its future constituents, the pilot polder authority strove to insert itself as an institution charting new state–society relations.

Due to its aura of difference, the pilot project was a great publicity stunt. Unfortunately, I missed the initial frenzy surrounding it, the effervescence of its initial meetings, and its celebratory inauguration at city hall. When I arrived on scene in 2014, the project was already way behind schedule, and some feared it was bound to fizzle out. Around that time, I accompanied Adin, one of SIMA's founding members, to a conference on resilience building organized by the NGO Mercy Corps. We shared a table with an employee of the organization and a delegate of Semarang's Badan Nasional Penanggulangan Bencana (National Disaster Management Agency). When the government representatives had finished their presentations on antiflooding measures, Adin was eager to make a comment. A man walked over to our table to hand Adin the microphone. He first introduced himself as a member of the SIMA agency. Apparently, not everybody knew what this abbreviation stood for, so he appended, "The agency that will run the polder, once it will have become reality [*kalau sudah jadi*]." This drew a chuckle from the audience, which was well aware of the project's massive delay. Despite a moment of embarrassment, Adin was able to quickly gather his thoughts and speak: "Disaster is our companion [*berkawan dengan bencana*], and yet the government has not committed to a concrete date by which flooding will be under control. It doesn't have the courage to promise that Semarang will be free of *rob*." The audience applauded moderately.

Adin's comment failed to conceal his disappointment with the project's delay. The initial momentum of the pilot project, after repeated corrections of implementation deadlines, had been difficult to sustain. Although the polder project had been touted as a special measure to solve tidal flooding, SIMA's members had come to realize that, as an unelected organization that was not directly involved in financing or building the polder infrastructure, it possessed no real power. Throughout the project's prolonged implementation phase, other temporalities eventually perturbed and undermined the board's vision of an alternative, participatory form of flood prevention. While the formation of the provisional, later officially decreed polder board played an important role in articulating the idea of flood governance through community involvement, its claim to flood governance was eventually superseded by other "temporalities of planning" (Abram 2014, 145). Over time, even the spatial changes brought about by the polder project turned out to be mere fixes, not that different from previous government attempts at controlling water flow. As such, the project had a special temporal relationship to politics. Its sudden eruption onto the political scene was allowed by what I call "pilot democracy." In a pilot democracy, crisis talk can allow for spontaneous political legitimacy. This legitimacy, however, flows from structures that can easily withdraw political support if they feel so compelled.

A reflexive anthropology of crisis allows us to understand the emergence and faltering of sociotechnical projects, such as the polder, that set out to manage ecological disaster. Crisis imaginaries serve to unsettle normalized structures and recondition expectations of the future. As Joseph Masco (2012, 1107) reminds us, "The concept of the extreme is relational, assuming a counterpoint to everyday experience marked regular, unexceptional, banal." The polder project hinged on blaming crisis on the "regular" way of doing things: the waxing and waning of drainage projects that fail to take seriously residents' imbrication in flooding as well as the stiffness of bureaucratic structures. Proponents of the project claimed to know that the solution to flooding lay in the responsibilization of locals through a water tax and increased political participation. These envisioned changes were quite extreme, and they allowed for exceptional developments. However, the infrastructure it sought to install was not sustainable in view of the floodplain's accelerated subsidence. While the polder project's ideas were certainly virtuous in the sense that they demonstrated interest in and empathy for riverside communities, its conceptual design did not allow for questioning the structural problems of urban development in the coastal plains of Semarang.

An Anthropology of (Flood) Crisis

The world over, climate change, often lumped together with global warming and sea level rise in media and activist language, is creating planetary ecological imbalances that will change the very way in which humans live and think (Klein 2014).[7] While the knowledge of a changing global climate has not radically changed the ways in which societies reproduce themselves (Prudham 2009; Swyngedouw 2010), visions of extreme ecological situations have already set in motion institutional processes and shaped subjectivities in culturally specific ways (Masco 2010). In Semarang, a diverse team was effectively put in charge of a drainage project that aimed to protect around eighty thousand inhabitants from tidal flooding. It interpreted *rob* as an event that, unless given immediate attention, would make the existing drainage system dysfunctional. This in turn would seriously endanger life in Semarang's North.

According to Janet Roitman (2013, 3), crisis narratives are critical assessments of events. Roitman argues that contemporary crises are constituted as objects of knowledge: speaking about crisis signals a specific knowledge of the world and the will to act on it. The very claim that we can act on history in fact depends on our conception of crisis. Roitman doesn't propose that crisis is a mere invention, an illusion. Rather, she offers an analysis of the social perception of crisis and how it relates to action. What does a departure from "normal time" entail for individuals and how do crisis narratives demarcate a given situation from perceived normality? I'm intrigued by the idea that the perception of crisis denotes a struggle with history and temporality. Let me concretize this by drawing a parallel, between flood infrastructure and the subprime market, which serves Roitman as an empirical example of the discursive effects of crisis. She points out that subprime mortgages—before they came to stand for everything that was wrong with the financial system—had been a standard practice for a long time. If, by analogy, tidal flooding is qualified as all that is wrong with Semarang's floodplain, how do we explain that it happened over more than a decade as a largely undisputed and even profitable thing? As I describe in chapter 3, a whole economy of endurance formed around tidal flooding. People and businesses prospered despite recurrent floods. Therefore, the polder project suggests a departure from a typical governmental approach, in view of the latter's "visible" inability to control *rob*. The perception of crisis that is attached to the polder project juxtaposes the "messy" time of northern flood adaptation to a new temporality. Henceforth, water management will be controlled, streamlined, and coordinated by a new institution that incorporates residents as responsible and rational actors.

In the case of the polder, a scenario of permanent flooding sought to legitimize a new kind of organization. Accordingly, the polder was meant to alleviate suffering and create proper conditions for economic growth. In turn, this envisioned economic growth was supposed to help deal with future ecological challenges, such as land subsidence and sea level rise. But what kind of future can an ideology of development imagine anyway? Walter Benjamin understood that capitalism entailed the (gradual) disappearance of radically different futures. Aware of the progressive disenchantment of the world by capitalist reproduction, Benjamin wanted to restore the possibility of the apocalypse to the oppressed, not necessarily in the form of a biblical deluge or nuclear annihilation but as the overthrowing of the capitalist elites. One needed to reenchant the world to see beyond the goal of universal equality pictured by capitalism (and never achieved). "The state of emergency, in which we live, is not the exception but the rule," according to Benjamin (1968, 257), whose last writings grappled with the horrendous outcome of German fascism. Benjamin could not accept the notion of a present that is not a transition. He therefore saw the necessity for a "real state of emergency." Eric Cazdyn (2007, 649) has pointed to the ubiquity with which disaster and crisis have been invoked in the post–Cold War era, while talk of revolution has actually become risqué—"not only rendered unspeakable but, more important . . . , unthinkable." While Cazdyn refers to the general absence of revolutionary talk in mainstream politics, talk of revolution has been dangerous terrain in Indonesia ever since the obliteration of Indonesia's once strong Communist Party, Partai Komunis Indonesia (PKI). Joshua Oppenheimer's documentary *The Look of Silence* (2013) and Eka Kurniawan's novel *Beauty Is a Wound* (2015; see Mrázek's [2016] review of the book) speak to the vacuum in Indonesia's public memory where there should otherwise be a critique of perpetrators and state abuses of power.

Masco (2017, S65) has theorized this shift by pointing to processes that led to the "collapsing of a more robust political sphere into the singular language of crisis." This political process allows for precise and pointed decision-making. Masco (2012, 1108) argues that crisis narratives that use the threat of catastrophe to justify institutional changes typically go hand in hand with a cultural formation devoted to the "normalization of violence (as war, as boom and bust capitalism, as environmental ruin)," such that it becomes the foundation of everyday life. Crisis discourse can therefore have profoundly nonprogressive consequences. I find this point crucial in light of Indonesia's current political situation. While imaginaries of political, social, or religious crisis symbolize a desire for the end of fear (of the Communist "threat," narcotics-affiliated crime, and political upheaval), it simultaneously normalizes the use of violence. In what follows, I argue that the polder outlined an end to crisis by normalizing specific emergency measures or forms of violence, such as evictions and the continuation of river normalization.

Any pilot project promises innovation and experimentation. Accordingly, the polder was framed in opposition to the status quo. The polder's key planning technology—resident participation—was supposed to open up a new future of economic betterment, democratic governance, and higher living standards. Importantly, the pilot project interpreted *rob* as a nuisance and a warning. Recurrent flooding signified that the area had reached a critical tipping point and would further deteriorate if nothing changed. Following Roitman (2013), I consider the polder project as relying on a crisis assessment insofar that it based potentially transformative action on knowledge of the future. According to the plan's logic, putting an end to flooding would entail economic growth, which in turn would improve local livelihoods. The flood crisis assessment was thus coupled with a specific future imaginary *and* knowledge of the interrelation of human economic behavior and environmental conditions.

However, as Roitman argues (2013), crisis discourse, despite arguing for a change of course, rarely has profoundly transformative effects. For instance, the portrayal of the 2008 financial crisis raised important questions as to the state of global capitalism. The bailout of defaulted banks suggested that although capitalism should be put in check, it was simply too profitable a system to let go. This raises important questions about the ways in which crisis discourse narrows our collective ability to imagine radically different futures.

Water Management the "Dutch Way"

In 2003 a technical agreement was signed between the city of Semarang and the Dutch Ministry of Transport, Public Works and Water Management. The agreement concerned the institutional steps toward water management through the use of an urban polder as a "sustainable solution for flooding problems."[8] As is typical for such agreements, it had a rather vaguely stated objective: to improve Semarang's existing water resources management. The agreement, however, did outline the creation of a financially autonomous polder authority modeled on Dutch examples. Unlike Indonesia, where the state and its local organs chiefly manage water and infrastructure, Dutch regional water authorities share powers. The technical agreement suggested embedding such a polder authority within the existing administrative apparatus in order to promote "democratic" water management. The project aimed at furthering Indonesia's decentralization and supporting the country's programs of community empowerment. A first feasibility study conducted between 2003 and 2004 led to the selection of East Semarang and the Banger River floodplain as a test site. According to Robin Peters (2012, 24),

> The waterfront city of Semarang, capital of central Java with 1.5 million inhabitants was put forward by the Indonesian government for the development of the polder concept . . . for four reasons; (1) the political landscape was considered to be less complex than the Jakarta region; (2) the city was considered to be representative for other Indonesian cities, such as Jakarta; (3) Semarang is known to be severely affected by floods on a daily base; and (4) the geographical location was convenient for most of the stakeholders.

Semarang's North stood out as a suitable project site by virtue of its extreme flood problems but also because it seemed less controversial, politically speaking. The repeated reference to Jakarta made it clear that, if successful, the pilot could be replicated in other Indonesian cities, especially the flood-plagued capital.[9] In 2007 another agreement was signed between HHSK, the municipal government of Semarang, and the Indonesian Ministry of Public Works to kick off the next "phase" of cooperation—the foundation of a provisional polder authority. From here on, the HHSK water engineer, Johan Helmer, took charge of the project on the Dutch end. Helmer was involved in the project for a considerable time and was a key actor in the polder project's development.

Notably, the polder project wasn't construed as typical "development aid." It was conceptualized as both a timely response to extreme flooding *and* a strategic investment. Following decades of unilateral development assistance, Indonesia is presently considered a potential client or paying patient that qualifies for Dutch water management therapy. In the new political climate of postcolonial emancipation characterized by economic partnerships in the private sector, the Dutch government was test-driving a cooperative development agenda coined "From Aid to Trade" (Van Marwijk Advies 2014). A helping hand from the former colonial "master" in the form of democratic water governance rhymed well with the logics of neoliberal global capitalism. Salvation had just been given a price tag, under the dubious justification that the Dutch government wanted to stop patronizing Indonesia. The considerable initial efforts of Dutch institutions to cofinance the project suggests that they considered this project an investment opportunity. Seizing it, however, meant finding common ground with the Indonesian partner on what exactly the problem was, as well as generating popular support for Dutch–Indonesian water expertise.

Framing Semarang's Flood Crisis

Helmer and I met multiple times during his visits to Semarang and in Rotterdam. Over dinner one evening, he told me about a crucial meeting he had with Semarang's mayor and the head of Bappeda's infrastructure unit. During the meeting, he had challenged the mayor, asking him outright if he really wanted change. Helmer had wanted to make it clear that he was not in Semarang "for fun," and that the condition of Semarang's floodplain demanded "urgent decision-making." In his view, bureaucrats who remained indifferent to tidal flooding were at fault, and their continued work, which ignored the catastrophe advancing upon the city, was a real threat. Helmer feared that politicians were unable to see the crisis for what it was, that Indonesian coastal cities were nearing a critical threshold, as expressed in increasingly regular tidal floods.[10] He knew from personal experience that sinking land poses a serious threat to both populations and the economy.

Helmer dedicated his preretirement years to the project, during which he visited Semarang countless times, kept a diary, posted small updates to Twitter, and lobbied the Dutch government repeatedly for support. But Helmer was not the only person pushing for change in view of the flood crisis. All the Dutch actors involved in the polder project hoped to "help Indonesia" get on top of the flooding issue. The Dutch water engineer Mondeel wanted to help implement the polder as soon as possible. For Mondeel, who often visited the project area and especially Kemijen, the crisis was visible: "You see some houses, they are wet all day. A lot of people [have] left the first floor and live on the second floor."

Scientifically assessing the crisis by conducting a feasibility study, including a topographic and economic survey, was an important step to generate initial financial support from the Indonesian and Dutch governments. However, as Mondeel recalled, the findings didn't immediately result in institutional support. Although he was able to showcase the scientific proof of the imminent danger of land subsidence compounded by sea level rise, the project didn't make it smoothly into the next planning phase. He complained that a lot of Indonesian politicians didn't want to commit publicly to the project. Instead, he said, "they want to stay below the radar, more or less. So, nobody really stands up and makes the project. . . . [Nobody] fight[s] for us. That's what you need to have a new prospect. That's difficult."

To make matters worse, in 2010 the Dutch government suddenly threatened to withdraw support. After years of trying to get the project off the ground, the Dutch government passed a law that prevented water authorities, such as the one that Helmer was working for, from financing international development projects. Helmer told me that "some Dutch taxpayers" were apparently upset that their water authorities were investing part of their resources abroad in times of domestic economic decline (and serious doubts as to the durability of the Netherlands' own dikes in view of sea level rise). The dream of a unique polder saving Semarang from drowning was suddenly in peril. But Helmer was adamant about continuing the project. After all, they had already "lost" money, that is, funding from the Dutch government, and his own board back in Rotterdam had contributed considerable staff resources to the project. At the end of the day, the Dutch public would be their judge. In addition, the Dutch government had clearly committed itself to this project and couldn't simply back out of it. Helmer said, "I went to foreign affairs and requested the continuation of the project. We have a memorandum of understanding. Do you really want to stop it?" In his opinion, his perseverance convinced the Dutch government to honor the agreements and consider the project as an investment in the Netherlands' future role in providing expertise to sinking cities worldwide. The project had Indonesian defenders as well. SIMA petitioned the Dutch foreign department for a continuation of the polder project. The board wrote a letter to the Dutch Minister for Development Cooperation in which the members praised the Dutch approach to flood control. It further argued that the project should not be interrupted, as the Indonesian government had taken note of this "new kind of organization" that was "supported by inhabitants and [the] business community in [the] Banger area."[11] SIMA thus played its own role in framing the project as indispensable to Semarang's future.

Semarang's municipality also cast the polder project as an urgent response to the tidal flooding that affected northern parts of the city. In 2014 the governor of Central Java, Ganjar Pranowo, stated that solving the "*rob* problem" (*permasalahan rob*) was a top priority of Semarang's administration—the city had to get this problem under control (*harus dapat diatasi*). He went on to compare the Banger Polder to a "pumping system" (*pola pompa*) that could "at least" work as a short-term solution (*solusi jangka pendek*) to flooding.[12] The vice president of Semarang's former water agency PSDA considered the polder project as a "part of their drainage master plan." In an interview, he expressed his belief that the polder system in question could help them get acute flooding under control. Importantly, he underlined that the polder infrastructure was supposed to handle the present situation. However, he nevertheless admitted that the polder would not safeguard the littoral: "In terms of a future without flooding, that's very difficult for us [to achieve], because we [the city] are located on the shore. In the future, unpredictable situations related [*dikaitkan*] to global warming will happen more often." Such statements from the Indonesian state regarding the polder suggest that the Dutch-designed polder never stood out as the only solution to Semarang's flooding problem. Rather, it was portrayed as a "pumping system" put to work in a delimited area of the city. In this sense, it was understood as a pressing but intermediate project. The polder was supposed to handle the area's present problems with tidal flooding, but uncertainty remained as to the shape and implementation of a long-term solution. While treating the situation as an emergency, the vice president of PSDA suggested that the city would stick to its drainage master plan after the polder had been implemented.

The intervention wasn't uncontested. Plans for the Banger polder triggered particularly strong public protest in the fishing neighborhood Tambak Lorok, as residents thought that construction of the dam would be disruptive to the bay's marine ecology. As specified in the design, the northern polder dam was going to close the channel just southeast of the settlement. Some residents argued that blocking bay water from entering the drainage tract would intensify tidal waves and result in ever-higher inundation levels. Given that numerous infrastructural interventions in the bay (e.g., the Marina Bay reclamation project) had left lasting ecological damage, including increased erosion and the destruction of fishing grounds, their apprehensions were warranted. These poor coastal communities have had to deal with tidal flooding largely on their own, and they had learned to use whatever political leverage they could muster to improve their situation. That is why they saw an opportunity for a tug of war with the municipality. While both the government and Dutch project partners considered the public protest as "unscientific," Helmer empathized with the fishermen. He remembered how at a public hearing, a resident warned them that his people would reject the polder construction if they did not stand to benefit from it. This local had complained that while *rob* had stopped or would be brought under control in other areas, there was still no change in his settlement.[13] Other residents threatened at the meeting to sabotage the polder infrastructure and take to the streets if their demands for appropriate flood protection remained unheard. The protest, in fact, proved successful, as the municipality saw itself compelled to commission a conceptual design for flood protection in Tambak Lorok.[14] The design, however, proved to be merely a lip service to the coastal communities of Tambak Lorok, given that the plan quickly disappeared into a drawer.[15] My point is that these protests, instead of unsettling Helmer's belief in the virtue of the polder project, underscored, at least in his eyes, its importance for the area's future safety. The *rob* crisis required that they start somewhere.

Roitman considers crisis narratives as second-order evaluations, assessments of a perceived situation that is specific to actors' positions in time and space. Positing a *rob* crisis forced the project partners to settle on a diagnostic of the present problem, a "questioning of the epistemological or ethical grounds of certain domains of life and thought" (Roitman

2013, 4). Both sides considered *rob* as indicative of a deep-reaching crisis of water infrastructure, yet this questioning served different purposes. To Semarang's government, the common narrative was a prerequisite for setting up new economic relations with the Netherlands and cloaking a significant state intervention in a participatory language. The Netherlands saw an opportunity to engage with Indonesia in a more emancipatory way while removing the scent of humanitarian aid from a postcolonial transfer of technology and knowledge.

To be sure, while the crisis diagnosis set in motion substantial institutional processes, it faced real resistance. When I started fieldwork in Semarang, the pumping house had already been constructed and gated off. Few people spoke to me about the extensive planning efforts involved in building the house. However, planning documents made it clear that the pumping house was controversial and required extensive moderation by the board's residential members. After all, the development saw to the eviction of about one hundred families and the demolishment of a neighborhood council building (*balai RW*).[16] To facilitate a conflict-free and clean land clearance process, a private consulting firm had been hired to produce a "social survey." Ostensibly, the survey was meant to count and categorize local residents, but it also seized information about their professional occupations, distance to work, and income levels. Residents received compensation based on how much money and material they had invested in their dwellings: concrete wall (*tembok*), partially brick (*1/2 bata*), or wood (*papan*). The last category was temporary (*temporer*), suggesting a form of nonmaterial existence. Residents living in "temporary" houses received the smallest compensations. Adin was involved in establishing initial contact between surveying agencies and residents so that the former could conduct the survey. According to him, people were more willing to listen to him, which helped prevent protests and allowed for "peaceful" evictions. However, he didn't know what had happened to the evictees afterward: where they currently lived and under what conditions. I couldn't find information on their new locations either. After this very personal, and I'm sure difficult, experience of facilitating eviction for the purpose of building the polder, he preferred not to be involved in such negotiations. In 2010, before the pumping house was constructed, a Climate Change Festival was organized on the now unoccupied parcel. According to the project leadership, residents had to become aware of the importance of this building, as it would become the central office of SIMA.

As a last example, I would like to mention that the construction of the pumping house also created significant tensions between the project partners. In particular, it revealed the limits of the common crisis diagnostic. After a strong rainfall, the building's freshly laid foundation suddenly appeared at risk because it had been built improperly. The construction error spiked conspiracy theories among Dutch project partners and created an awareness of the wider politics of infrastructure provision in Indonesia. Helmer requested an immediate inspection of the incident, as he suspected fraudulent use of building materials. As he explained it to me, he "ordered" the mayor to suspend construction immediately so that they could inspect the building during the Dutch delegation's next visit. However, when he arrived in Semarang, concrete had been poured over the evidence. It seems as if this purging act, in a literal and symbolic sense, allowed both sides to ignore the deeply political terrain of infrastructural modernization in Indonesia and move on with the project.

The Polder Board

In 2010, shortly after the polder design release, an official polder board was created based on a memorandum of understanding and a mayoral decree (Perwal) that made the Dutch water authority (HHSK) and the municipal government of Semarang primary partners in the project.[17]

> There was a ceremony in April 2010 at which the Mayor of Semarang, the Governor of the Central Java province, and the Secretary General of the Ministry of Public Works signed treaties for transferring powers to the water [polder] board—some of which are far-reaching by Indonesian standards. The water board [*sic*] will devise, write, and introduce all rules, procedures, and protocols necessary to ensure the proper working of the polder. (Van der Pal 2012, 24)

As previously mentioned, the name SIMA is created from the first syllable of the Dutch board's geographical base and the second syllable of Semarang. Vice chairman Sumarmo liked to point out that SIMA's name was also reminiscent of Shima, the female monarch of Kalingga, an ancient kingdom on the northern coast of Central Java. The queen is celebrated for her unmerciful sincerity, reflected in a strictly enforced law against thievery introduced during her reign. This underlining of the polder board's deeply Indonesian identity and integrity kept surfacing in the board's self-representation. The polder authority wanted to emulate "democratic" Dutch water governance while promoting a distinctly Javanese style of political participation in operation and maintenance of the water infrastructure. SIMA thus represented an enactment of traditional values and simultaneously integrated core ideas of Dutch governance. The polder authority worked hard to silence postcolonial tensions and smooth over a potentially controversial transfer of foreign expertise.

To that end, SIMA consisted of governmental and nongovernmental members. I focus on the latter group, as they were the most vocal defenders of a resident-run polder. The board's chairman, Suseno Darsono, is a professor of water resources planning and management at the Department of Civil Engineering, Diponegoro University.[18] He also worked as a consultant for Semarang's government, conducting hydrological research and designing drainage policies. I often saw him speaking with the director of Bappeda's infrastructural unit. As the most senior and highly educated member, he was often received by officials in private, while SIMA's residential members only got to exchange with them at official meetings. SIMA's Technical, Operation and Management Division was coordinated by another water expert, Imam Wahyudi, who teaches civil engineering at Semarang's Sultan Agung Islamic University. He co-supervised research projects carried out by Dutch interns. The board's legal expert was the most active member with government ties. He often attended internal meetings as well as events. The board further consisted of two representatives of Semarang's business owners, a secretary, and two people working in administrative support. The secretary, Indah, was a very important member of SIMA.[19] No event escaped her knowledge. She kept track of deadlines, meetings, and expenditures. As a former resident of the Banger floodplain and daughter of lower-class parents, she strongly sympathized with the project and entertained affectionate friendships with the senior resident members, Sumarmo and Adin. On a normal day, she would be the first at city hall, going through the newspapers to collect information relating to water infrastructure. I frequently arrived second and gossiped with Indah over a cup of instant coffee.

The foundational members of SIMA were not randomly selected. Rather, they were recruited by the Dutch and the government in light of their capacities or previous involvement in water-related issues. In 2006 the idea to involve residents in water management was making the rounds in Semarang. The then mayor Sukawi Sutarip wished to involve residents in surveys covering a range of river maintenance indicators. Sukawi thought that flooding occurred because the existing drainage system—consisting of ten subsystems—was no longer functioning properly. He launched a task force comprising local academics, government staff, and representatives of the Local Empowerment Boards for Neighborhoods (LPMK). According to Sukawi, kampung residents were holders of "meticulous knowledge" (*ilmu titen*) that they developed by virtue of their daily engagement with water. Sukawi further described residents as deeply familiar with the history of the drainage system. As a result, he reasoned that they should be involved in the development of improvement plans.

Adin decided to join the initiative. He signed up for a Tim Khusus Subsistem (Special Subsystem Team) that was formed for his neighborhood, Kemijen. His contemporary, Sumarmo, was also intrigued by the opportunity. Sumarmo had had a short political career as a member of Semarang's parliament (Dewan Perwakilan Rakyat Daerah Kota Semarang) in the 1980s. During my fieldwork, I had the honor to live with Sumarmo's family for five months. Sumarmo proved to be not only a well-connected *tokoh masyarakat* (local figure) and a skilled entrepreneur but also a passionate karaoke enthusiast and loving father. I was on good terms with many members of the team, but rarely or never visited their homes. Conversely, it was not uncommon for me to wake to the sound of Sumarmo's melodic humming first thing in the morning. He and I often discussed local as well as national politics (for instance, Indonesia's controversial execution of nine foreigners convicted of drug trafficking in April 2015) over late-night cups of coffee. I learned a lot from him about some of the darkest events in Semarang's history, such as when Suharto's henchmen rounded up and executed alleged communists in the 1960s and blacklisted "gangsters" in the 1980s.[20] As he often pointed out, a reformed Indonesia opened doors for community organizers like him—Javanese men who are respected by their communities and who dared to speak out against the powerful.

His participation in the subsystem survey marked the beginning of his education in drainage technology. As subsystem members, they collected river samples to evaluate water quality and helped measure water depth and sedimentation levels. The subsystem was a precursor to the polder board in that it increased residential involvement in water governance. Both programs saw in engaged residents the missing link between the local population and infrastructure provision. Whenever the subject was breached, Sumarmo always had a lot to say about the Special Subsystem Team. He called it a participatory government program that hired residents to compile data of drainage channels in their respective neighborhood subsystem. The data was to be communicated to local leaders (*sosialisasikan*) and used to help inform improvement plans. Sumarmo, like all other resident participants, was offered a small amount of monetary compensation for his work. All survey activities were supposed to be concluded in only one week—a laughable prospect, according to Sumarmo. He judged that the hired residential assistants were bound to do a sloppy job in view of the short window of time and low remuneration. Sumarmo also criticized the program because of its inconsequentiality. According to him, the results of their survey were never disseminated. He concluded that they had been "used" (*diperalat*) by the administration. In his view, the creation of the subsystem teams had merely given the illusion of citizen participation. In 2007, when the government invited Sumarmo to participate in a "short course" on drainage technology offered by Dutch drainage experts, he decided to take it, being attracted by the opportunity to learn from foreign specialists. Through this course, both Sumarmo and Adin met the Dutch delegation behind the polder project. Along with a handful of other subsystem participants, they were eventually approached to help build a polder authority.

In the following pages, I show why Adin and Sumarmo embraced their new positions as members of the polder board. They not only represented the Banger floodplain communities, certainly a prestigious role, but also had an opportunity to help reform governance structures. Joshua Barker, Erik Harms, and Johan Lindquist (2013, 166) argue that urban ethnography "is not a question of finding figures that are somehow representative of a social group or place, but rather of understanding particular figures as evocative nodes that reveal relationships and forms of mediation between individual lives and wider social processes." I see the polder board as just this kind of evocative node, a point in the lives of Sumarmo and Adin at which their individual paths intersect with Semarang's infrastructural crisis and democratization. The project endowed them with meaningful and lasting roles in urban affairs that didn't exist for residents in previous government initiatives. Both tried to use this role to remake the kampung in a fashion that jibed with their political ideologies and desires for the floodplain communities' future. As such, their involvement in this project reveals the social hierarchies at play in the becoming and failure of Banger Polder.

Deploying the Local Population

SIMA enthused and thrilled many, especially elite advocates of participatory planning and bottom-up development within the Indonesian government and abroad. A report heralded the polder project as built purely on civil participation (*murni partisipasi masyarakat*) and not requiring much governmental funding (*tidak membebani anggaran*). Two successive directors of Bappeda's infrastructural unit lauded the project, while another considered it particularly promising. Members of Himpunan Ahli Teknik Hidraulik Indonesia (Indonesia's Association of Hydraulic Engineers) hoped that the polder board would become a "vessel" (*wadah*) of civil participation in the fight against flooding and termed the system the "most appropriate" (*paling cocok*) of solutions (Darsono and Mestika 2011, 313). The following conversation between the Dutch project worker Roy and Pak Timbul, a resident of Kemijen, was recorded at the beginning of a meeting that took place at SIMA's office in April 2015. One of SIMA's mandates was to train residents in hydrology and infrastructural maintenance. A casual conversation between Roy and Timbul shows how the dominant framework of these meetings portrayed participants as victims by virtue of the extent of the tidal flooding they endured and their local status.

> ROY: You're all locals? [*Semua asli?*]
> TIMBUL: All residents. [*Semua penduduk.*]
> ROY: You guys often flooded? [*Sering banjir?*]
> TIMBUL: Primarily tidal flooding. [*Pertama rob.*]

ROY: [It happens] almost every day? [*Hampir setiap hari?*]
TIMBUL: Not that often. [*Ngak juga.*]

Semarang's government also saw the project as an opportunity, as its participatory design promised to attract international attention and financial assistance from the central government for infrastructural maintenance and river modernization. The project shared many characteristics with what Jamie Peck (2014, 396; see also Harvey 1989) has termed "entrepreneurial urbanism"—the result of the competitive conditions of neoliberalism "within which . . . entrepreneurial urban strategies have been formulated."

The Indonesian Kementerian Pekerjaan Umum dan Perumahan Rakyat (PUPR, Ministry of Public Works and Housing) stated in a 2019 report that the "function and position" of the polder authority was of utmost importance in supporting the polder because this unit would play "the role of managing the daily operation and maintenance of the system." Although some form of public participation is integrated into most Indonesian development projects, whether at the conception or implementation stage, the polder was portrayed as an institution with an unprecedented level and quality of civil participation. Such a level of civil participation was in fact barely thinkable for many government workers I interviewed. Employees of the local water agency, PSDA, admitted that they had virtually "no clue" what SIMA would be doing once the dams, pumps, and riverbanks were constructed. Would ordinary residents run and maintain the pumps? While SIMA was often heralded as a promising institution capable of saving Semarang (and why not Indonesia?) from flooding, the participatory hype around the project also elicited confusion and doubt. Although financial independence seemed to be a shared goal in view of budgetary constraints, legal experts within the government voiced concern about the board's legal compatibility with municipal regulations. Would the board have the necessary sovereignty to levy taxes? Can ordinary residents work in the service of the state? Although staffed with water and legal experts, the polder board was often unable to assuage these reservations and doubts.

One solution was to invoke long-standing kampung traditions. The board's turn to kampung traditions would not surprise scholars of Indonesian development politics. Most famously, the World Bank started using a community approach in 2000 (see Li 2007). Furthermore, as John Sullivan's (1986) research in Java has shown, the Indonesian state has long extracted labor from local populations by calling on "traditional" values, such as the concept of mutual assistance, or gotong-royong. While Sullivan's findings depict these state–society relations in rural Java, Jan Newberry (2008) has made similar observations for urban kampungs. The polder project's idea to levy a local tax traditionally called *iuran* (as explained in the case of pumping communities in Kemijen in chapter 2) to run the polder was not that new. What raised eyebrows was the idea of funneling this tax money to an autonomous authority that would independently run state-financed drainage infrastructure. To popularize this outlandish idea, SIMA started publishing monthly leaflets about the polder project and established a publicly accessible library consisting of informational brochures and books on water management. It also welcomed delegations of employees working at different levels of the Indonesian government who wanted to learn about this new project. In addition to its own weekly meetings, it also hosted regular open-door events where at least half of the board's members were present. Staff of the central government came to visit the birthplace of the pilot project, which by that point had even started making the news in the Netherlands.[21] SIMA's members also witnessed the regular coming and going of national and foreign, mostly Dutch, delegations. With some regularity, groups of Dutch undergraduate students would spend a few weeks in Semarang on Dutch grants in order to conduct research on aspects of the project as part of their final exam or research internship. I was often considered an "expert source" by these students because I spoke Indonesian and I was familiar with the locations and demographic of the area where the polder's main organs (pumping house, retention basin, and dam) were going to be built. As an "observer" of the project, I also regularly accompanied SIMA's members to public events and site inspections.

 Two years after SIMA's launch, construction of the pumping house finally began. This was when SIMA began a series of public hearings in the designated polder territory. Soon, or so they thought at the time, their organization's office would move into the pumping house. They wanted to at least prepare the local population for the presence of this new institution. As members in charge of public relations, Adin and Sumarmo thus promoted the polder project in the affected neighborhoods and "tested" their readiness to play a supportive role in its realization. According to the board's view, planners and engineers lacked privileged access and insight into the lives of poor inhabitants of the polder territory. The "residential members" thus functioned as a gateway to the polder territory's social consciousness—as crucial anchors of social networking and cultural translators. Sumarmo and Adin regularly helped organize consultation meetings (*sosialisasi*) at which they "informed" residents about the polder design and operation mode. Between Sumarmo and Adin, it was often Sumarmo who addressed larger audiences and spoke up during meetings with officials. Sumarmo was at ease with audiences and always had a good joke up his sleeve. He even composed a special song to educate the public concerning proper garbage disposal.

The "Polder" Coming Apart

The polder board, in particular its residential members, was expected to "socialize" the project design before operation. It was also key in educating the population about the project's objectives. At times, this meant imbuing an outwardly unspectacular water infrastructure with an entirely new meaning. When I started fieldwork in 2014, all the pumps and interior outfitting of SIMA's pumping house were still missing. Three years after construction work had begun, the pumping station still had no apparent function. Nobody used it. Goats grazed undisturbed in its sun-drenched front yard. Helmer noted in his travel log that, at one point, "inside the barbwire fences around the building of the pumping station, a veritable small fairground had been erected. The area is serving as a local venue, as long as the prestigious office of our water board has not risen." The building was meant to be a formidable metaphor for the change that the Dutch–Indonesian polder promised to bring: "dry feet for all"—the coming of a *new time* that included all riverbank communities. For many years, however, it remained empty, an optimism on pause and overgrown.

In Indonesia, pending infrastructure projects attract the attention of the feared Komisi Pemberantasan Korupsi (KPK, Corruption Eradication Commission). For this reason, the building's liminal status posed a real problem to Semarang's government. Rumors circulated that the suspended polder project had fallen prey to corruption. Why else would work have come to a stop? Worse yet, the polder project was tainted by the arrest of one of its more influential defenders. The deputy head of Semarang's water agency was charged with corruption following investigations of another water infrastructure project.[22] In 2014, in the face of growing public discontent about the project's delay, the incumbent mayor announced that Semarang's government would expedite completion of the polder by dipping into its own annual budget. After all, what good was an empty pumping house? And yet the rumors never fully dissipated. For example, my landlord was convinced that the mayor had simply parked the project funds in a bank account in order to accumulate interest. The delay seemed torturous to the Dutch partners, who feared that these rumors might implicate them. When a contractor finally began preparing the station for the arrival of the pumps, Adin felt compelled to visit the pumping house to check on the installation process. I tagged along.

The contractor in charge welcomed our visit as an opportunity to rant about Indonesian water management. According to him, water governance failed at the hands of miscalculating, fraudulent dilettantes. The design for the polder was no different to him. He, for instance, liked to use cheap American pumps, ordered with "MWI Pumps" based out of Florida, which provided fast and low-priced delivery, while the government's choice of pumps had to take into consideration various regulations and contractual obligations. As the contractor went on about the pumps he ordered, assembled, and installed "all over Indonesia," the uniqueness of the Banger Polder seemed to dissipate before us. I noticed Adin's slight defensiveness. Instead of inaugurating a new time, one that kampung residents could inhabit in a meaningful way, the pumps themselves turned out to be handmaiden to much more mundane goals: profit, speed, and convenience. Even the design turned out to be squarely derivative. From looking at the design sheets, he was quickly able to deduce that an Indonesian bureaucrat had made the technical sketches, not a Dutch engineer. Furthermore, he also pointed out that the shape of the pumping house looked like an expedient "copy-paste" job: "all over Indonesia they are using the same architectural style [for pumping houses]." "How funny," he added, that although there was no actual regulation for the design of pumping houses, they all looked uniform (*seragam*). The pumping house's copycat design compromised Adin's vision of Banger Polder as a unique, democratic experiment, an experiment he had helped evict scores of residents for. In this brief encounter, the provision of pumps, enshrined as constitutive of a new relationship between water and local governance, was revealed to be subject to the whims of powerful politicians.

The renewed sense of urgency created by persistent rumors of corruption fostered the conditions for urgent and undemocratic measures. While being central to the polder, the pumps were themselves dependent on other components. In order to function properly, the pumping station had to be balanced by adequate water storage. The second major intervention thus concerned retaining water. According to the Dutch design, successful flood protection necessitated the acquisition of inhabited land for water retention. The dam in the estuary promised to keep tidal water out but also trapped runoff. During the wet season, not even the strongest pumps could handle the onslaught of rainwater. The excess water therefore had to be retained somewhere. While the construction of the pumping station alone required evicting dozens of households, the retention basin turned out to be the bigger and costlier engineering challenge. Existing regulations and mixed land tenure made constructing the basin tricky. The Indonesian Railways Company (PT KAI) owned the patch of land slated for construction. Accordingly, Semarang's master plan indicates the parcel in question as part of the transportation sector.[23] Yet, despite being counted as a "national asset," the land had gradually become an integral part of Kemijen, a home to hundreds of residents.[24] Although living on the land was deemed illegal at the time, several hundred families had houses there, and the subdistrict government had been obligingly taxing these "squatters." The first inhabitants were arguably employees of PT KAI, who acquired building permits but were not allowed to purchase land titles. Initially, the company had toyed with the idea of developing a container depot on the land. This would probably have necessitated draining the land, large parts of which are flooded and used for fish farming. The pond was in some ways a kin to the forest ruins described by Anna Lowenhaupt Tsing (2015) in *The Mushroom at the End of the World*; it was a place that had come to sit uneasily between regulations, patchy law enforcement, and capitalist logics yet was nevertheless productive for marginalized populations.

Following Mondeel's recommendation (Mondeel and Budinetro 2010), the government pursued the possibility of building a retention basin here, and PT KAI eventually agreed to lease the land to the government in 2011. Just like that, the polder construction turned this degraded land into a valuable asset. Like good "salvage" capitalists, as Tsing might say, the communities around the fishpond started vying for compensation as victims of eviction. As a result, when more and more residents started signaling demands for compensation, and more and more houses started popping up on the project plot in anticipation of imagined recompense, the polder construction stalled once again. At this point, the government saw the need to properly negotiate with the community. SIMA did not reach out to the community to socialize the polder construction plans, as it had with the pumping house. In this instance, the evictions were considered a governmental affair.

Contrary to what Jane Guyer (2007) and others have argued (see introduction), Simone Abram (2014, 130) has shown that planning doesn't necessarily evacuate the future through apolitical decision-making processes that obey profit margins and capital reproduction. In fact, she contends, contemporary forms of planning continue to both envisage long-term futures *and* attempt to address immediate issues. Yet what planning rarely considers is the near and medium past. This temporality gets evacuated from the practices of planning experts and bureaucrats whose "relentless focus on worlds yet to be and work yet to be done" easily ignores meaningful relations with the past. In the case of the polder, expediting construction meant abandoning the near past of this land. The encounter with this unruly past affected the planning process, leading to the destabilization of "planning's selective technologies of moving through time" (Abram 2014, 130). As shown earlier in this chapter, the pumps depended on the retention capacity of the basin. Both hydrological components of the polder were, in turn, part of a vision in which water was properly drained and channeled to allow for economic progress. Yet spatiotemporal restructuring efforts brought to the fore issues that imperiled its future-making plans.

Notably, it was the sense of urgency perpetuated by the Dutch partners' pressing and mounting suspicions of corruption that helped the project convalesce after verging repeatedly on failure. The Dutch polder scheme imagined a helpless subject that was surviving, as it were, in the wet and productive milieu of places out of step with progress. That subject, in turn, justifies the continuation of modernizing efforts, no matter the cost. The polder project also critically rests on a culturally specific image of the "swamp." In it, only a morally crippled version of the human can survive, waiting for its chance to modernize. The polder is neither a complete return to modernist teleology nor a postdevelopmentalist scheme. It holds on to visions of modernization but portrays residential actors as the essential link between environmental crisis and a "modern" future.

A "Sad Monument"

The delay in construction of the polder project put strong political pressure on the city's office to act. The government risked coming across as unable to step up to the plate at a moment that it had helped frame as a crisis. Despite or because of sudden efforts to finalize the project, the heart piece of the polder—the "prestigious polder office" within the pumping house—threatened to become a "sad monument," as one of the polder experts described it in a conversation. As an empty building without function, the pumping house ironized public infrastructure and questioned the efficacy of Semarang's city planning.

This final section of this chapter focuses on the eventual cooptation of the polder project. In particular, I show that SIMA's residential members were facing (and still face) power hierarchies that limit their ability to play an influential role in transforming flood governance. Adin, in particular, feared that the project would ultimately fall prey to the very system it tried to transform. Sumarmo, too, expressed uneasiness about the project's sudden progress. But he worried more about SIMA losing status as a legitimate authority. Previously, he had been exuberant to be called upon to organize visits to the polder construction site because he could show off his knowledge of water infrastructure and the polder. Despite having a modest education—"*cuman SMA*" (equivalent to a high school diploma)—people were "impressed" by his civilized manners (*heran saya budaya*) on such occasions. At some point, they even asked him—a small-time kampung resident, he emphasized to me—whether he thought the Dutch trustworthy, as if they considered him a diplomat in the employ of the government. Quite diplomatically, he would answer that the project did not solely consist of Dutch people but also involved many local academics (*ilmuwan*) and residents. But when we brought a few Dutch engineering undergrads to the pumping house one day, he confessed that he was embarrassed by the state of the project. "This is all I can show them," he told me, referring to the empty pumping house. At the very least, the installation of the pumps meant that the polder was materializing and that they hadn't made empty promises to the public.

Adin's worries related to the government's rash reaction to the delay of the retention basin construction. As public pressure on the government to conclude the project grew, rushed implementation became more and more probable. He also expressed concern about the municipal water agency PSDA as it became increasingly impatient with the project, especially its ebbing participatory ambitions. When the polder seemed to come back from the dead because of the government's financial step-in, SIMA expressed grave concerns that the original institutional design of the polder would be revamped into "normal" flood governance. Instead of being met with institutional restructuring, they feared the polder would be assimilated by a government agency when in fact SIMA was meant to retain many prerogatives, such as choosing its own staff.

In April 2015, I arrived late to an important meeting at SIMA. I was surprised to see four people dressed in typical government batik shirts sitting at the table. During the meeting, Adin spoke at length about the importance of recruiting staff and doing on-the-job training while the pumps were being installed. He complained to the government representatives that they had not prepared contracts for staffing SIMA's technical positions. Sumarmo chimed in, stating that they also had to start recruiting staff to train for the polder's daily operation. The government representatives, especially a PSDA employee, seemed uncooperative. They wanted to delay recruitment until all infrastructural components had been paid for. SIMA's legal expert intervened: "SIMA isn't some wild idea [*ide liar*] but the result of a cooperation between the Netherlands and Indonesia. Our actions have to aim at creating an institution with its own authority [*diharapkan punya otoritas sendiri*]." He added that what they hoped to adopt from the Dutch wasn't just some technical aspect of drainage but an institutional innovation (*ke lembaganya*).

Sumarmo was frustrated by the outcome of the meeting and the government's apparent hesitation to take SIMA seriously. I asked him why they couldn't criticize the government publicly for letting them down. He said that this kind of rough talk (*bicara keras*) was only possible in meetings since speaking badly about the involved stakeholders to the press would only mean harming themselves (*dipukul sendiri*). As a government initiative, SIMA was not only highly dependent on the priorities of the involved government agencies; it was also seen as complicit in controversies surrounding the evictions of residents.

Around the time that the government seemed set to proceed with evictions, Adin and I went for a walk through the neighborhood and stopped to sit on the Banger River embankment for a while. We contemplated the neighborhood's fishponds, a stagnant body of opaque water. In 2009, when the city government officially released the design for Polder Banger, the fishponds were slated to become a retention basin with a catchment surface of nine hectares.[25] "I have many friends in the settlement," Adin said. I asked him who was going to compensate the evictees' losses in terms of property (*pengantian rugi*) and who would carry out the evictions (*bongkaran*). He answered: "This is a government affair. I don't want this to be a conflict between the people. It ought to be a government job [*seharusnya pemerintah*]." The polder development had introduced a social fault line. Yet Adin's own social network suggested that social space in the North wasn't as neatly divisible. His wife's cousin, Pak Rianto, who was also one of Adin's collaborators in the resident organization Komayu, filed a complaint with the government because his house was probably going to be removed. Adin liked to express that he was speaking and acting in the name of residents like Arief and Deni—poor kampung dwellers whose lives seemed stuck somewhere between ecological, economic, and medical crisis. He wanted "dry feet" for *all* residents. Yet the realization of the polder forced him to draw a line between people with and without official housing documents. Naturally, the situation caused an "inner conflict" (*konflik batin*) in Adin. The polder project had suddenly become synonymous with any other top-down development project he had witnessed in the past. Furthermore, while Adin supported a "democratic" and peaceful resolution of the land conflict, the government's obscure course of action made him think that the project would materialize without effecting a real democratization of water governance.

By the time I had to wrap up my research project, Adin and Sumarmo were clearly losing hope that the polder they had envisioned would ever become reality. After I visited the pumping station one day, I decided to swing by Adin's house. I could immediately sense that he was in a bad mood and stopped asking research questions, hoping he would soon launch into a polyglot discussion on philosophy as he often liked to do. But he remained silent. We both yawned and sank deeper into our chairs. At one point, I dared to ask why he was so tired. Although he did not seem delighted by the topic, he answered: "Because of this . . . because it's not happening. It's exhausting for the morale [*melemah semangatnya*]; it's hard to stay motivated." When someone came to visit the project site, he always had to pull himself up and put on a positive face. He patted his chest: "It's embarrassing when people come here to learn but find nothing. My motivation is jittering [*redup*], like a lamp without enough current." Naively, I asked him why he continued his work. He explained, "I'm just flowing [*mengalir saja*]. I've got no choice." The polder, as the epitome of democratic water management, had become a constraint for Adin. Instead of helping him and SIMA undo the social hierarchies established on the swamp to date, it put him in an awkward situation in which he had to make empty promises.

The "Already Flooded" and the New Chronic

When I returned to Semarang in September 2018, I was impatient to find out if *rob* was still an issue. I had heard from friends that the area was dry and safe from tidal flooding after the polder began operating. I had moved back for two months in order to catch up with recent events and learn about the social and physical effects of the infrastructure project. Upon arrival, I could immediately see that something had changed. The water level of the Banger River had dropped by about one meter. A provisional dam was preventing tidewater from causing the

Banger River to overflow, while pumps were transferring water from the Banger River to the East Flood Canal. *Rob* seemed under control.

Given this meaningful step toward flood safety, I was very surprised when I learned that SIMA had been kicked out of its old office. According to Adin, SIMA had been "kindly evicted" (*diusir secara halus*). The old office had been a spacious room located in the mayoral building of Semarang's city hall. Here, SIMA's board members had regularly received delegations from other Indonesian cities: engineers, government workers, and foreign students. The big office had seemed commensurate with SIMA's unique goal of establishing Indonesia's first community-run drainage system. Now, however, it had been relocated to an office in the main building of the city hall, on the same floor as Semarang's Planning Agency, Bappeda. The new office was a windowless dungeon too small to fit all of SIMA's board members. In addition, the board members mostly convened after working hours, when most state officials had long ended their working day and city hall was locked up. As a result, SIMA barely used the new office. But it hadn't dispersed, either. Since the move, SIMA's board members had begun to hold meetings at the private residences of its members. During my visit, the team decided to gather in Sumarmo's driveway.

As I helped Sumarmo lay out a large red synthetic carpet for guests to sit on, I silently wondered why they were meeting at all. Hadn't they reached their goal to provide "dry feet for all"? Once everybody was settled onto the ground, instead of jubilation, I noted concern and frustration upon everyone's faces. Although the polder had been able to stop flooding, the board members consider it unfinished because the project had failed to fundamentally change the way water management is done in Semarang. The board, established in 2007, still possessed no legal authority regarding any aspect of the operation of the polder. In fact, the group felt cut out from the project, and getting evicted from the prestigious old office had made this impression even stronger. The government was walking away from the project even though it wasn't complete. SIMA was now looking to partner with local universities in order to meet the necessary legal requirements to achieve the status of a corporation (*badan hukum*), a prerequisite for taking action in the name of the polder project.

At the meeting, it was decided that SIMA would independently organize a bike tour on polder territory to popularize its agenda and remind the population of the benefits and possibilities of a resident-run Polder Banger authority. After the event, which failed to attract much attention, Sumarmo and I had a long conversation about the bike event and the status of the project. During this conversation, he strongly criticized how the bike ride, which produced a minor turnout, had been disturbed by "other activities" (*terpotong-potong kegiatan lain*). In fact, the *camat* (district head) who attended the event used it to promote his own political agenda. As a result, they didn't get a chance to advertise the polder and were visibly sidelined by the *camat*. We also spoke about the apparent reluctance of the municipal government to actually finish the project. Some opponents, according to Sumarmo, thought that it was mere advertisement (*iklan*) for the Dutch, just merchandise (*jualan*). While he understood their critique, he underlined the mutual benefits of this deal: the kind of water management to which they aspired, that they needed, didn't exist in Indonesia. In addition to government inaction, he pointed to obscure interests (*oknum*) that were undermining the project. After all, Sumarmo surmised, the real polder would render a whole range of informal deals virtually unnecessary. The polder would erase a reliable source of income for people who profited from the current remedial form of flood prevention. Peninggian was too lucrative to let go.

Cazdyn (2012, 5) argues that the "chronic" is about a "certain relation to time, a relation to the present and the future, as well as the capacity to shape these temporal realities." This chronic mode insists on "maintaining the system and perpetually managing its constitutive crisis, rather than confronting . . . the system's own death" (5). In the chronic, the future is spelled out in advance, "granting to the meantime an impossible location that is heading somewhere and nowhere at once" (4). In this chapter, I showed that while the polder project tried to unsettle and inaugurate a new time, it hinged on reproducing social, political, and institutional relations. The polder was originally conceptualized to alleviate and not remedy flooding—rather, it was aimed at carving out time for the floodplain inhabitants to become politically involved in the improvement of the area. According to its design, galloping land subsidence and predicted sea level rise would ensure that the polder would stop functioning properly in fifteen years. Not only Helmer but also the former vice president of Semarang's water agency had acknowledged this. In this sense, the polder project postponed a decision on the future of the floodplain. It assumed that the polder infrastructure would provide the necessary conditions for bringing economic progress to the region, thereby proposing a scenario in which other, less actionable factors would play an immensely important role. As such, it only contained the seeds for long-term safety from floods, without actually effecting structural change. The cases of Sumarmo and Adin capture an individual experience of chronic time that is perhaps best expressed by Adin's metaphor of "just flowing," which poetically refers to his relationship to time.

Despite catchy labels, such as "pilot" or "participative governance," the flood prevention project ended up repeating well-rehearsed government practices, such as oblique evictions and paternalistic planning. Participation by residents in the polder scheme merely helped give credibility to an assessment of a crisis, define the technical contours of a solution, and develop a project able to fix water flow and educate citizens. Yet after fixing this problem, the drainage project returned to its normal course of implementation. As Cazdyn (2012) argues, although the chronic dictates management-style solutions for complex socioecological problems, and so encourages tame public participation while suppressing real alternatives, it evidently produces criticisms at the same time. That critique is echoed by SIMA's residential members. It rests on long-standing democratic aspirations as well as local imaginaries of a drowning floodplain. In the end, however, the polder project helped inaugurate a discourse that considers recurrent crisis as constitutive of everyday life, a discourse that does not accept the death of the normative drainage system. After all, it is the drainage system's failure that provides lucrative business to some of the floodplain's inhabitants.

Masco (2012, 1107) has pointed out that lifeworlds today are "founded on machines" with specific logics and affects. As normalized infrastructural scaffoldings of the everyday, they are further "extreme, in the sense of being unprecedentedly violent" (1107). In such worlds, a "highly developed social commitment to normalizing extremes" (1107) is required to maintain the illusion of progress that modern political theory elevates to both the end of history and its means. Masco considers progress as "a perpetual engine of improving the infrastructures of everyday life as well as the morality of those living within it" (1114). While the polder project destabilized governmental practices, it also normalized infrastructural adaptation to tidal flooding based on the latter's presumed adverse effects on economic and national progress. The polder conjures up a crisis in order to make a historically plausible case for institutional change. Nonetheless, the polder project helped install a "new set of fantasies and short circuits that prevent reflexive critique" (1115). It forced residents to work toward an impossible future, one in which they are always "already flooded."

AFTERWORD

When I returned to Kemijen in 2018, I was pleased to see that certain infrastructural conditions had improved. The government had made good on its promise to complete and start operating the pumping house in North Kemijen. Since 2017, when pumps first began diverting water from the Banger River to the adjacent East Flood Canal, the area had been widely kept safe from tidal floods. Most residents confirmed the relief from flooding in their daily lives, and the streets stayed dry during high tide. As a matter of fact, the deputy chief of Bappeda had checked the area off his list. Meanwhile, Johan Helmer, the Dutch water engineer who most fervently had fought for the realization of the polder system, had retired. The powerful people behind the project had "done their job," as it were.

On my first day back in the "field," Adin and I went to check on the water level of the Banger River. The river had in fact dropped by a meter and a half. The fleet of pumps and the provisional dam had done their designated technical jobs. Adin pointed out that on some days you could even glimpse the river floor. Low water levels enabled other infrastructures: all bridges across the river were now suspended above the water surface again instead of sinking into the brown muck. The polder had turned the Banger into a cesspool, an engineered receptacle of wastewaters and runoff. The slimy ground exuded the familiar foul smell, suggesting that the Banger wouldn't need a new name. To Adin and many of his friends, the polder meant immediate and tangible improvement, a much-desired relief from *rob* for the masses.

I was particularly elated when I saw that Ariel's and Arief's situation had also somewhat improved. When I arrived in Semarang for a short visit, I found Arief outside his home, painting newly built concrete flower pots. His family had received a pro-poor grant to elevate parts of their home and the veranda. Arief had always been skeptical of government aid, but he accepted the support as a token of recognition for his work. Although the government had recently refused their proposal to fix the gutter in his street, Arief was still pleased with the improvement of their domestic situation. The polder was keeping the river level in check, but they still had to run their pump every day to keep the gutters from clogging. Arief also explained that some residents still complained about *rob*. For instance, houses located south of the northern artery still experienced flooding. Also, one resident informed me that the neighborhood's remaining fishponds now overflowed at night. Outside the polder's hydrological limits, *rob* continued inexorably. The coastal neighborhood Tambak Rejo struggled with increasingly frequent and strong tidal inundations (Ley 2021), while water pooled for months in streets of segments of the poor subdistrict Kali Gawe. *Rob* had also ultimately made landfall in Indah's neighborhood. She was distressed by the home improvement works in her home. Although they had bought their house in 2015, they were already forced to raise both foundation and roof, a considerable financial strain on her family's budget. Because SIMA had been kicked out of its original office, I went to see her on the seventh floor of City Hall, where Bappeda has its offices. She was embarrassed to work here among "actual" officials with contracts and pension plans because the polder project didn't grant her any long-term occupational status. She felt like an impostor. Some days later, she broke out in tears when she informed me about her boss's derogatory remarks about SIMA. She recalled a situation where he bluntly disparaged the residential members of her team. She had invested so much time in the project and was now helplessly witnessing how it all slowly came undone. After ten years of tiresome meetings, hopeful rallying, and imaginatively working toward participatory water management, the idea merely fizzled out.

In conversations with Sumarmo, still deputy chairman of the Banger Polder board, I was privy to a similar perspective on the polder project. While Sumarmo recognized that the polder represented positive change for residents, he was very unhappy with the results. In fact, he admitted to me that he was embarrassed when people asked him about it. To him, it was "still not working." Even from a technical side, he considered the project incomplete and sloppy (*tidak konsekuen*). In fact, the dam hadn't been built according to the technical design, and the retention basin was missing. This partial implementation risked suggesting a false safety when in fact sudden surges in precipitation could easily overburden the hydrological functioning of the polder. But what bothered Sumarmo most was that the social concept of the polder had not taken root. Here, he was referring to the clean culture (*budaya bersih*) concept that the polder was supposed to symbolize. According to him, "only one in thirty is actually inspired" by it.

Hearing these premonitions, I knew that my own fears had come true. Bappeda and the Dutch project team had effectively walked away from an incomplete system that replaced a fragmentary but tested infrastructural network of residential self-help devices with a temporary technical fix. Perhaps even worse, this fix short-circuited the democratic processes that were supposed to politically transform water management in Semarang's North. Providing "dry feet" was supposed to create a time frame for fundamental changes in urban governance, changes that would help democratize water management. The government's expedited execution, however, reveals the power of the chronic according to which the government just manages flood risk to offer some sort of stability. While it promised a different future, the polder project looped residents into a perpetual present. They will have to continue to build on borrowed time, as the polder project expires in less than fifteen years.

This book covers a vast period of time. I begin in late colonial Semarang, describing the emergence of coastal kampungs on the northeastern border of the city. Based on the meager historical record of these poor kampungs located along rivers—which functioned as sewers and irrigation systems at the same time—I reconstruct the ways in which the colonial administration pursued its "enlightened" style of urban governance. More specifically, I elucidate how it tried to address human problems in the rapidly urbanizing port area of Semarang, which overflowed with destitute indigenous newcomers due to rural dispossession and an urban racial hierarchy. Drawing on this colonial legacy, the postindependence state carried on with a similar treatment of these "problem zones" by repathologizing kampungs of Semarang's North, that is, addressing certain issues, such as crime and a lack of hygiene, while ignoring other problems—critical infrastructure, impoverishment. Following this, I employed a different lens to produce a finer, close-up view of these places' infrastructural transformation under the influence of political and spatial "normalization." Naturally, the image this produced is more complex, and a variety of temporalities become apparent. An ethnography of the present, divided into acts, illustrated what looks like a never-ending loop of time in the lives of riverside dwellers. The infrastructural ruins of river modernization produce an everlasting present in which residents must constantly mobilize materials, capital, and physical efforts to repair and thus maintain a barely functional landscape of water, and therefore human life. I thus juxtaposed the developmentalist time of normalization as a tool of state making to the chronic present that it produces—a nesting of temporalities—which layers quasi-events and leads nowhere. Peninggian and the baroque disposition—two forms of building on borrowed time (I'm sure there are more)—are development *reductio ad absurdum*. Enter the Dutch–Indonesian polder project that wants to transform urban governance but only serves to further normalization after local flood prevention has been brought more firmly under the auspices of the state.

I have ethnographically elucidated a situation of chronic breakdown that escapes the diagnostics of this state. While chronic breakdown easily disappears from the horizons of policy makers, this situation deserves our attention as anthropologists. It is not only an ethnographic reality, but it shapes, even cripples, the ability of individuals to engage with the plans that governments currently devise to address climate change. The possibility of breakdown spellbinds residents and sucks their efforts and resources, while their work only extends the dissatisfying present. This meantime, according to Eric Cazdyn (2012), is a dominant function of the chronic. The meantime allows institutions to develop a government of time in line with the shifting conditions of neoliberal capitalism. It forces urban residents to make temporary arrangements (Simone 2020) that do not stick for very long, while killing the drive for meaningful political change. It further raises the question of whether Semarang's floodplain residents are the "already dead" of urbanization, which in conditions of contemporary capitalism condemns marginalized inhabitants of coastal areas to a life with chronic infrastructural breakdown.

I'm not a friend of conclusions. It feels presumptuous of me to try to reach a final word on anything, especially if it is as contingent as an urban delta and its manifold forms of existence. I don't dare to make future predictions based on the material I have collected. But there are a few things I would like to say upon reexamining the claims that I have made in this book. In a book focused on water in Southeast Asia, Peter Boomgaard (2007) provides an overview of types and representations of water in this vast and politically and culturally diverse region. Boomgaard does a respectable job at presenting anecdotes, data, and ethnography culled from a plethora of studies and historical material. Regarding Indonesia, Boomgaard specifically points to the dangers of living by the sea due to regular exposure to disease, winds and floods, and the problems of river pollution. He also mentions the dangers of global warming, which constitutes a new challenge to coastal populations. As the present work has revealed, since at least the beginning of the twentieth century, living in the swampy North exposed dwellers to many dangers, in particular outbreaks of disease and seasonal flooding. Over time, due to the stigmatization of the North, danger also started to emerge from within the kampung, as gangsters and rogue state officials roamed this landscape and often clashed, leaving a trail of death and fear. Thus, the water of the Banger River became constructed as dark, dangerous, and demoralizing.

Today, in the coastal kampungs of Semarang, danger arises from dysfunctional infrastructure that was supposed to cure the North of its assumed "dark" characteristics. Instead of the state offering a cure, communities and individuals must mobilize their own resources in hopes of adapting to an unequally distributed threat—*rob*. Residents claim that *rob* began after the normalization of riverbanks in the wake of state forays into the dark realm of northern kampungs. *Rob* became a normalized occurrence, whose effects on peoples' health and levels of agency were unaccounted for and chalked up to the vagaries of globalization and ecological change. Boomgaard, in fact, hints at threats that are difficult to account for, such as poor-quality drinking water. This danger, a result of underinvested or abandoned infrastructures, such as the flood gates of the *jalan inspeksi,* riverbanks, and drainage channels, however, "does not yet show up in the overall mortality figures" (2007, 14). *Rob,* as well, has corrosive effects that do not show in official figures. Though well intended and timely, studies of land subsidence and climate change cannot account for such accumulated and combined risks. Andrea Ballestero (2019) has recently shown how difficult it is to understand the workings of aquifers, that is, how and why the underground soaks up, redistributes, or ceases to produce water. How do we begin to delineate such vast underground infrastructures? We know that an important factor in tidal flooding, land subsidence, occurs due to groundwater extraction, but how do we account for the effects of nested, decaying infrastructures? What are the ethics of such an undertaking?

Rob as the aftermath of normalization is like Cazdyn's (2012) bombs that force people to metaphorically live under the shadow of destruction. Ethnographic dwelling, a methodological "sinking in," has allowed me to illustrate the consequences of building on borrowed time; these consequences accumulate (building) effects and stem from attempts at outlasting and enduring the decomposing present. Kim Fortun (2013, n.p.) points to the "renewed relevance of ethnography and anthropology in these—disastrous—times." In this era, so she argues, "disaster can be chronic as well as acute. Asthmatic, or delivered by a toxic cloud that kills on contact."

As this book has argued, the chronic is an effect of political marginalization and produces hopeful strategies to deal with the dead ends of infrastructural adaptation. The strategies respond to and foster a twisting meantime. I suspect that we will see more of such "meantimes" in the future. What liberal states today—stripped of their tax bases and dismantled by neoliberal policies—can protect the millions, perhaps billions, of humans who are living in zones endangered by climate change around the world, where chaos is never moving into order but simply into new forms of chaos? Climate change is here, and yet most governments are still pretending that it is a distant threat, one that can be dealt with in due time. As I have shown, this meantime is not just a deadly trap for many subjects and communities who must respond to the pressing onset of ecological transformations but also a political dead end that deprives people of a dignified and more self-determined future.

ACKNOWLEDGMENTS

There were many radiant stars on the path that led to this book. I must first thank the people who cared to plant the thought in me that this research needed a broader audience: Tania Li and Joshua Barker's unwavering belief in this project especially helped see it through. Tania, who is to be in part credited for the title of this book, always had an open ear and an open door. Two young and superhuman heroines of this discipline, Katie Kilroy-Marac and Naisargi Dave, also encouraged me to publish, which meant a lot to me. Abidin Kusno provided insightful comments on a first draft of the book and was an inspiring influence throughout my research in Indonesia.

This book would not exist without the support and knowledge of people I met in Indonesia. Fieldwork stays before and after PhD research were facilitated and enriched by an untiring Pujo Semedi, whose sustained support, home-roasted coffee, and interest in my work gave me strength during tough stretches. In Semarang, I was lucky to always be surrounded by friends, mentors, and family. Eny helped me make first contacts, find shelter, and offered welcome distraction from my day job by recruiting me as a durian pancake vendor. Pak Puji became a mentor, guide, and reliable source of laughter. His staunch grassroots organizing and fearless habit of talking truth to power with all the fervor of a young activist infected his friends in Kemijen and me. The Sumono family gave me a place to live and find respite. Their trust in me and the constant joy and bustle of their home allowed me to relax after long days at city hall. I admit that I played with the idea of writing an ethnography of their curious, labyrinthic house, a haven to visitors and remote family members, instead of addressing flooding. My friend Wahyu helped me dock on to SIMA, had me over for delicious Soto, and thankfully introduced me to one of Semarang's biggest poets, Djawahir Muhammad. The SIMA board, its hard-working members, and its most fervent Dutch supporters, Johan Helmer and Roy Kraft van Ermel, always made me feel welcome. Swinging by their (air-conditioned) office was one of my favorite things to do. The radiant Pak Purnomo at Bappeda tolerated my incessant questions and allowed me to get an on-the-job perspective on urban governance. Ibu Ita, Pak Safrinal, Pak Luthfi, and others were incredibly gracious and patient "colleagues" during my time at Bappeda. I thoroughly enjoyed spending time and marveling at the chaotic nature of governance with Meimei, Ratri, Roni, and the Mercy Corps Indonesia crew. I strongly benefited from the knowledge and innovative work of the Kolektif Hysteria and enjoyed hanging with Adin, Purna, and Arief. Claudia Stoicescu was a wonderful host whenever I passed through Jakarta. It was a delight to spend time with Pak Agus and his generous family.

I thank Lauren Yapp from the bottom of my heart for being a true companion in fieldwork life and beyond.

When it comes to writing books, I'm still a novice. I'm deeply indebted to Jesse Henderson for bearing with my writerly ambitions and helping me improve the prose and structure of this book. Jesse's editing skills are only surpassed by his grasp of poetic language. He managed to walk the fine line between getting the tone just right and preserving my own ESL literary quirks.

Then, there are those who read and commented on very convoluted first versions of chapters in this book and deserve praise: Lauren Yapp, Shireen Kashmeri, Salvatore Giusto, Maria Martika, Letha Victor, Joey Youssef, Hollis Moore, Lisa Davidson, Michael Lambek, Sabine Mohammed, Sina Emde, and Catherine Scheer. Franz Krause deserves a special mention for showing me how to work toward publication with gusto. The following people helped clarify core arguments of this book with either lucid questions or candid critique: Abidin Kusno, Tania Li, Joshua Barker, Ursula Rao, Scott Prudham, Shiho Tsatsuka, Sheri Gibbings, Lea Stepan, Eli Elinoff, and Tyson Vaughan. Many others provided thoughtful comments on papers at conferences and workshops and I can't remember all of them, but they too shaped this book. Golschied Rahmani helped with copyediting while I taught in Heidelberg, when time was of the essence.

One anonymous reviewer and Jan Newberry read various versions of the book manuscript and were extremely generous with their wisdom and time. Their careful reading and commenting improved this book in many ways. I'm very grateful for the energy they invested in this project. I also thank Pieter Martin, editor at the University of Minnesota Press, for taking the time to understand my research material and argument. He persistently helped navigate through a challenging reviewing and editing process. At the Press, Anne Wrenn and Rachel Moeller kindly provided logistical support.

The cover of this book showcases not only the finesse and fearlessness with which residents of Semarang brave infrastructural breakdown in their daily lives but also Aji Styawan's keen photographic eye for the city's struggle with tidal flooding. Although Aji and I never met in person, he trusted in my work and let me use one of his masterpieces.

Sheila McMahon made my own copyediting skills look extremely amateur. While ironing out the inconsistencies of my writing, her friendliness certainly helped see this project through during the homestretch.

Getting here was not an easy feat. I was lucky to be part of a tremendously resilient group of students at the University of Toronto's Department of Anthropology. While workload, expectations, and financial struggles regularly tipped work-life balances and made academia seem like an exercise of treading water, they not only braved a psychologically toxic environment but also infused everyday campus life with art, humor, and compassion. I especially thank Sardar Saadi (and Gülay Kilicaslan), Sophia Jaworksi, Emily Gilbert, Jesse Henderson, Bronwyn Frey, Emily Hertzman, Laura Beach, Jean Chia, Jessika Tremblay, Stephen Campbell, Shayne Dahl, and George Mantzios. In Toronto, my roommate Dukhee Nam took me in when I was in a rough patch. Without Alex and Aaron Marques, this book would have never seen the light of day.

Lastly, I want to thank my parents, Petra and Heinz Ley, who not only allowed but encouraged me to pursue my dreams in and through academia, a mysterious world they will, sadly, never understand. My brother, Martin Ley, was and continues to be a beacon of support.

I cannot thank Lea enough for consistently telling me to chill out.

GLOSSARY

air pasang: High tide
ambrol: Collapse
banjir: Flooding, mainly when caused by rain
Bappeda: Regional Development Planning Agency
bencana: Natural catastrophe
BKM (Badan Keswadayaan Masyarakat): Community-level organization, especially its board of trustees
BLH (Badan Lingkunan Hidup): Indonesian Provincial Environmental Agency
bocor: Porous
dewan: Council or body consisting of a limited number of people
drainase: Canalization, drainage
faskel (fasilitator kelurahan): Person facilitating between subdistrict government and BKM
gotong-royong: Mutual cooperation
HHSK (Hoogheemraadschap van Schieland en de Krimpenerwaard): Dutch regional water authority in Schieland en de Krimpenerwaard
iuran: Local tax system managed by residents
jalan inspeksi: Inspection road located on riverbanks that separate buildings from body of water

Jalan Pengapon: Pengapon Street
Jalan Santai: Communal walking event
kali: River
Kali Banger: Banger River
kampung: Popular neighborhood
kecamatan: City district, political unit
kelurahan: Subdistrict, subdistrict office
kerja bakti: Communal work at neighborhood level
keturunan tanah: Land subsidence
kota: City
Kota Lama: Old Town in Northern Semarang with remnants of Dutch architecture
KSB (Kelompok Siaga Bencana): Disaster Preparedness Group
lurah: Head of subdistrict government
meluap: Overflow
Musrenbang (Musyawarah Perencanaan Pembangunan): Community-informed budgeting process supposed to articulate residential needs to local government
normalisasi sungai: Government policy banning housing from riverbanks and standardizing river maintenance
oknum: Person or being with negative intentions
pavingisasi: Road- and street-paving activities
pembangunan: Building, synonymous with development
pemerintah: Government
peninggian, meninggikan: Elevation, elevate, or raise
Perda (Peraturan Daerah) Drainase: Regional Drainage Policy
perkembangan: Development
pesisir: Coastal area
pintu air: Water gate
PLPBK (Program Penataan Lingkungan Permukiman Berbasis Komunitas): Community-based environmental zoning and settlement regulation program, part of World Bank participatory development scheme
PNPM (Program Nasional Pemberdayaan Masyarakat): National community-driven and empowerment program launched in 2007
pompanisasi: Water pump association
preman: Gangster

PSDA (Pengelolaan Sumber Daya Air): Water Resources Management
PT KAI: Indonesian Railway Company
rakyat: Little people
rob: Tidal flooding, tidewater
RT (Rukun Tetangga): Neighborhood unit
ruwatan: Cleansing ritual freeing people or beings from bad luck
RW (Rukun Warga): Community-level administrative unit
saluran: Small canal, ditch
sosialisasi: Government-organized event where project goals and procedures are communicated to local residents
sungai: River (Indonesian)
talut: Slope, riverbank
tanah uruk: Gravel or dirt used in landfill
tanggul sungai: River embankment
Tim Khusus Subsistem: State-salaried team of local residents tasked with surveying river maintenance
tokoh masyarakat: Local authority
uruk: Landfill
warung: Small shop selling beverages and food

NOTES

Introduction

Many thanks to my former colleague George Mantzios for introducing me to Theodoros Chiotis's poem "Perfusion."

1. I have used pseudonyms for individuals quoted in this book, unless they are public figures or consented to having their identity revealed.

2. I noticed how unwelcome the slum label was when speaking with residents about the recent increase in registered human immunodeficiency virus (HIV) cases in Semarang. While illnesses related to water supply, waste disposal, and garbage are feared, they are considered "normal" aspects of kampung life. Digestive diseases, moreover, can be treated since antibiotics have become available to most residents regardless of income. People wish to "ignore" HIV because it is a highly stigmatized disease. Most residents refuse to educate themselves about it and disregard the calls from neighborhood leaders to get tested, as the virus is associated with impure and morally degraded places—the port and its illegal settlements.

3. Bowen acknowledges that gotong-royong existed before independence. However, he argues that the Indonesian state relied in unprecedented ways on the exaction of affective and unpaid material labor.

4. This is an incomplete portrayal of kampung communities throughout Indonesia. Some kampungs are more fractured and transient than Sullivan suggested in this seminal piece. Kampungs can be divided along ethnic and religious lines. Kampung communities were also instrumentalized as watchdogs

for the Indonesian state, which resulted in lasting conflicts and tensions between residents.

5. "Community Driven Development in Indonesia," World Bank, August 1, 2014, http://www.worldbank.org/en/country/indonesia/brief/community-driven-development-in-indonesia.

6. For cosmological interpretations of water and the sea in Indonesia, see Trumbull 2013. A historical overview of water in Southeast Asia is provided by Boomgaard 2007.

1. Becoming

1. Parwito, "Ganjar akan bangun tanggul raksasa atasi banjir di Semarang," *Merdeka,* February 20, 2014, https://www.merdeka.com/peristiwa/ganjar-akan-bangun-tanggul-raksasa-atasi-banjir-di-semarang.html.

2. Jaringan Dokumentasi dan Informasi Hukum Kota Semarang, "Perda 7 Tahun 2014 Tentang Rencana Induk Sistem Drainase 2011-2031," September 26, 2014, http://jdih.semarangkota.go.id/ildis_v2/public/pencarian/53/detail.

3. In the Dutch East Indies, Malay served the Dutch as a surrogate language of governance, "because very few non-Dutch spoke their language" (Siegel 1997, 14). As such, Malay was a language of authority and lacked a cultural basis for its non-Dutch users. The language most commonly used in North Semarang is Javanese.

4. For this research, I spent one week in Rotterdam's Het Nieuwe Instituut (The New Institute), whose collection contains sketches, correspondence, and articles from Dutch architects and town planners who resided or worked in Semarang. I also had access to the Koninklijk Instituut voor Taal-, Land- en Volkenkunde (KITLV, Royal Netherlands Institute of Southeast Asian and Caribbean Studies) archive in Leiden for another week in order to collect further historical data on North Semarang.

5. See, for example, "Schets Van De Verbinding van de Boven- en Benedenstad van Semarang," a 1918 map by Dutch architect H. Maclaine Pont (1884–1971).

6. The municipalities and the central government quarreled for more than twenty years over the legal framing of interventions into indigenous affairs, while people succumbed to recurrent epidemics and starved in the kampungs (Cobban 1974). Though some scholars have lauded Semarang's council as ahead of its time, the situation must have felt quite different for the indigenous population—more like a protracted state of legal exception, the end of which would be

decided by Dutch democrats arguing introspectively over the colonial agenda, fair budgets, and subsidies many thousands of miles away.

7. Map of annual mortality rates in Semarang, 1894–1904, reprinted in Tillema 1913.

2. Stuck

1. Based on a jingle well known to the residents of Surabaya, Anton Lucas and Ariel Djati (2000) have argued that it was common not only to use rivers as a dumping ground for rubbish and industrial waste but also to dispose of dead animals, such as dogs. In the 1980s, as a result of new environmental regulations, dumping rubbish, or the dead, was outlawed in Surabaya.

2. The LPMK is a sister organization of the BKM (see introduction). It replaced its predecessor, the Lembaga Ketahanan Masyarakat Desa (LKMD, Council for Village Security), in 2007 in the wake of decentralization regulations. See "Peraturan Daerah Kabupaten," no. 13, 2006.

3. Today, this fear often has a class character, since kampungs are primarily home to the urban proletariat and the socially disenfranchised.

4. Reviewing academic literature on colonial town planning in Semarang, I found a persistent representation of the kampung as chaotic and disorderly. In his economic-historical work, Cobban represents the kampung as an unconsolidated, unorganized mass that "could not have done much more than contribute in an *inchoate* way to the general desire for change in the conditions of their existence under colonial rule" (1988, 287, my emphasis). The "inchoateness" of kampung communities is a trope that continues to animate Indonesian urban governance, as exemplified by community participation programs and bottom-up initiatives involving mapping and social restructuring. While the "disorderly" kampung is a trapping of Dutch rule and racial politics, its legacy is powerful and continues to be glimpsed in scientific literature on the kampung.

5. Daniel Garr (1989) showed that KIP was driven by a humanitarian ethic and plain economics: "in some kelurahan [subdistricts], kampung self-help efforts constituted as much as 91% of the local share of KIP resources, virtually eclipsing the role of the municipality." In Semarang's southern neighbor, Solo, "nearly 60% of KIP funding consisted of contributions from within the assisted kampung. And in Semarang, an important seaport to the north, nearly two-thirds of KIP project costs came from this source." Garr concludes that while KIP selection criteria were initially biased toward deficiencies in physical conditions, they became rather skewed to the magnitude of the local contribution. Deficiencies in sanitary conditions determined whether kampungs

were eligible for support in Semarang or Surakarta. However, the "worst first" criterion was mostly outweighed by economic expedience (Garr 1989, 80). In Semarang, only three kampungs were selected for improvement works: Bandarharjo, Bugangan, and Kuningan (Direktorat Tata Kota dan Tata Daerah 1989). None of the northeastern kampungs received direct attention. Though comprehensive in theory, the state's effort to "regularize the legal status of popular settlements and homeownership" (Kusno 2012, 33) actually contributed to polarizing urban space, improving the infrastructure of strategically located areas while postponing improvement in less central ones. In view of KIP's strong reliance on community labor, it is no wonder that the Northeast was not conceived of as a candidate for successful kampung upgrading. In the 1970s, this area struggled with high crime and poverty rates due to advanced marginalization and thus probably didn't count as a potential candidate for improvement.

6. "Kali Banger 'Diruwat,' Warga Minta Normalisasi Dilanjutkan," *Suara Merdeka*, accessed January 25, 2017, http://www.suaramerdeka.com/harian/0505/03/kot09.html.

3. Floating

I want to express my gratitude to the poet Djawahir Muhammad for allowing me to quote his work. Further, I wouldn't have come across his writings if not for my good friend and interlocutor Wahyu Ambari.

1. "Building with Nature Indonesia: Securing Eroding Delta Coastlines," Wetlands International, July 17, 2014, https://www.wetlands.org/video/building-with-nature-indonesia-securing-eroding-delta-coastlines/.

2. EcoShape is ostensibly guided by a new hydraulic engineering principle called "building with nature," an approach whose aim is to simultaneously boost nature, recreation, and the economy.

3. The minimum wage was IDR 1,685,000 per month in 2015, which equals USD 117.

4. Urban ethnographers have recognized the heuristic value of walking through the city. As Sudhir Venkatesh noted recently (2015, 347), urban ethnographers often promise to bring the reader into the "invisible or closed-off social worlds" of their research subjects. The kampung is not a closed-off social world, but its inhabitants often don't like to speak about their financial strategies. Exploring the city alongside locals can be a suitable technique for revealing the differential abilities of neighbors to deal with hardship. Adin lives on the street portrayed in these photographs, and we often strolled through the neighborhood. I consider walking an everyday practice in Kemijen through

which urban space is made (de Certeau 1984). By taking photographs from the same angle in the same, fairly level street, I tried to bring some formality to this descriptive method. At the same time, Adin and I would speak about the houses, assessing their viability in the face of *rob*. Alternatively, contemplating a particular house could elicit comments from other locals. Monica Montserrat Degen and Gillian Rose (2012, 3281) contend that "walk-alongs" can "produce richly evocative sensory impressions." Furthermore, as Helena Holgersson (2016, 223) has argued, when applied in ethnographic research, "a walk-along plays out more as a dialogue than an inquiry." The ethnographer and the research participant can engage in a speculative dialogue that reveals diverse contextual knowledge of spaces and their functionality. Walk-alongs allow the ethnographer to experience the material and imaginary making of the city firsthand.

4. Figuring

1. After meeting with the lurah, Wahyu had to return home as quickly as possible to oversee construction work on his house. He had hired workers to raise the floors of the bathroom and kitchen. In other words, when he referred to people's urgent and lasting needs, he was also referring to his own.

2. The PLPBK works through the BKM in providing infrastructural development. In cooperation with an assigned PNPM facilitator (the so-called *faskel* or *fasilitator kelurahan*), the BKM has to fulfill specific criteria during the period of funding, such as being audited once a year by a public accountant and publishing an annual economic report. Further, the BKM must host an annual citizen's conference. I spoke with the *faskel* about Kemijen's performance in the program since the creation of the BKM in 2000. He generally lauded their participation but also noted that they struggled to reach the official quorum at their last organizational meeting.

3. "Civil society" is a concept that plays a central role in neoliberal thought: "'decentralization' has become, along with 'civil society,' 'social capital' and 'good governance,' an integral part of the contemporary neo-institutionalist lexicon, especially of the aspects which are intended to draw greater attention to 'social' development" (Hadiz 2004, 700).

4. According to Aspinall (2013), the authoritarian regime of Suharto had aspired to base its rule on the silent consent of the masses but was unable to cultivate such a public. However, it succeeded in silencing specific Others, such as Chinese Indonesian minorities living in Java and other regions as well as government critics with more socialist agendas. Such Others have been under more or less violent attack since the 1960s.

5. I counted more people in line at the gas station—to get a last "sip" of cheap fuel before the president's decree came into force—than in front of city hall.

6. "Legislation focused on decentralization to the third tier of government, namely regencies (*kabupaten*) and cities (*kota*), with the result that the role of provincial governments in managing the coordination of inter-local affairs has been severely attenuated in the era of decentralization" (Bunnell et al. 2013, 859).

7. Siegel (1993) has noted for Java that people sensed the political weakness of the social hierarchy during Suharto's reign. He observed, for instance, that residents of a poor area established neighborhood watches (*ronda*) whose task was to police the boundaries of the local community. The use of violence toward outsiders (thieves and/or Chinese Indonesians) was a result of suppressed anxieties surrounding fragmentation and dissolution. Siegel concluded that fear was *localized* and "made into hierarchy" (38).

8. The government itself was quite busy deflecting and incorporating such criticism during my fieldwork. The salience of tidal flooding in public discourse forced the government to produce and control knowledge about *rob*. For example, *rob* was often spatialized; that is, government reports and surveys tried to map incidents of *rob* and trace its impacts in regional terms, glossing over finer details and differences. Furthermore, in reports and newspaper articles, experts invoked several reasons for tidal flooding, highlighting illegal groundwater extraction, lack of infrastructure, and littering. Given the vague contours of *rob*, the government often promised in roundabout ways to measure up to the challenge of building "resilience" to flooding. For instance, Semarang's mayor said he wanted to improve the city's "physical" and "non-physical" urban infrastructure (Semarang Municipality 2016, 7) in hopes of building a "modern" and "developing" city. Dominating the spatial outline and discourse of *rob* seemed, however, a daunting task, as alleged triggers rarely led straightforwardly to effects.

9. The multiple triggers of local land subsidence (Marfai and King 2008) and its geographic variability make it tricky to identify vulnerable neighborhoods, in geological terms.

10. While the head of Karang Taruna has recently argued that the organization should overcome its image of local government bellboy (*pelayan kantor kelurahan*), its goals often still correspond to state interests, as recently seen in the state-backed "war on drugs" (*pemberantasan narkoba*)—one of Karang Taruna's latest projects.

11. The numbers are based on the subdistrict's unpublished 2015 infrastructure priority list ("Daftar Skala Prioritas Pembangunan Sarana Prasarana Lingkungan Pemukiman Yang Dibiayai Dari APBD 2015").

12. The following chapter discusses the rumors of eviction in more detail.

13. Repeated conversations with planners have revealed that climate is considered a minor factor in the production of *rob*.

5. Promise

1. The five *K*s in Penta K Labs stand for *kamu/kita* (you/us), *kelas/kampus* (class/campus), *komunitas* (community), kampung, and *kota* (city).

2. In a symbolic attempt to fuse the two countries in the context of the project, SIMA combines Schieland—the name of the partnering water authority's home region—and the second syllable of Semarang.

3. Kraft van Ermel expressed these views in my interviews with him. Tellingly, Jarred Diamond (2005) has elevated the Dutch polder to an exemplary political tool for governing global ecological and economic processes.

4. The Dutch water authority HHSK shouldered the operational costs of SIMA and expenses for design (albeit with support from numerous Dutch institutions, as mentioned previously).

5. This expected result of the polder construction was never communicated in public hearings during the time of my fieldwork, suggesting that accelerated land subsidence was scientifically acceptable but did not translate smoothly into social acceptance.

6. In the Netherlands, citizens pay annual taxes for the management of regional water systems, such as maintenance of dikes and control of water levels. This water tax does not account for drinking water usage, which is monitored and charged separately by water suppliers (see Hendriks 2001; Woldendorp and Keman 2007).

7. Before the world climate summit in Paris in 2015, the pope warned that "it would be sad—and I dare say even *catastrophic*—were particular interests to prevail over the common good and lead to manipulating information in order to protect their own plans and projects" (Al Jazeera 2015, my emphasis).

8. The technical agreement concerned the "Institutional Strengthening of Water Management in an Urban Polder System as a Sustainable Solution for Flooding Problems" and was signed by Semarang's Agency for Research and Development (Ministry of Settlements and Regional Infrastructure) and the Dutch Ministry of Transport, Public Works and Water Management on February 18, 2003.

9. Peters (2012, 25) notes that the feasibility study was not followed by immediate action, as "it took the Dutch some time to find funds for continuation of the project." Furthermore, the mayor of Semarang supported a rival

project. Only when this project was quashed did the municipality again show interest in the Dutch idea of a polder based on parity among stakeholders.

10. While I interviewed Johan Helmer multiple times in both Rotterdam and Semarang, I learned much about his views from informal conversations during meetings or trips.

11. "Continuation of Banger Pilot Polder," letter from SIMA Polder Board to Prof. Dr. H. P. M. Knapen, Dutch Minister for Development Cooperation, January 4, 2011.

12. Provincial Government of Central Java, "Pemprov Pantau Pembangunan Polder Banger," May 19, 2014, http://jatengprov.go.id/id/berita-utama/pemprov-pantau-pembangunan-polder-banger (no longer available online).

13. "Polder Banger Tidak Kunjung Selesai, Ini Masalahnya," *Tribun Jateng*, May 19, 2014, https://jateng.tribunnews.com/2014/05/19/polder-banger-tidak-kunjung-selesai-ini-masalahnya.

14. HHSK in cooperation with Witteveen+Bos, "Basis of Design and Conceptual Design Report," 2010, in author's possession.

15. The concept had a comeback five years later, after the president of Indonesia, Jokowi, paid a visit to the drowning fishing community.

16. One survey was conducted by the Jakarta-based consulting firm PT Arkonin: "PRA LARAP—Pekerjaan Pembangunan Rumah Pompa Polder Kali Banger Semarang," Jakarta, 2009, in author's possession.

17. Until 2016, the Dutch Embassy in Jakarta covered the bulk of SIMA's administrative costs, while it did not commit to any expenses for physical components of the polder. The group of Indonesian and Dutch actors involved in the project has undergone several mutations that I do not account for here. Since 2006, HHSK has been the most important institutional partner on the Dutch side, and the Rotterdam-based water authority continues to be deeply involved in the project.

18. Suseno received his PhD from Colorado University.

19. The first secretary was fired by Helmer, after the board found out that she had embezzled funds.

20. Once, for example, men came to Sumarmo's house and asked him to drive a truck convoy of vigilantes through his neighborhood.

21. "Beschrijving HHSK Banger Polder Pilot Project," RTL Nieuws, December 5 2012; see also "RTL nieuws 5 december 2012 Semarang," YouTube, May 18, 2014, https://www.youtube.com/watch?v=dz3QnwLSkfk.

22. "KASUS KORUPSI: Sekretaris Dinas PSDA Kota Semarang Ditahan Kejaksaan," *Solopos*, August 27, 2015, https://www.solopos.com/kasus-korupsi-sekretaris-dinas-psda-kota-semarang-ditahan-kejaksaan-636954.

23. Pemerintah Kota Semarang, "Rencana Tata Ruang Wilayah Kota Semarang Tahun 2011–2031," http://bappeda.semarangkota.go.id/v2/wp-content/uploads/2012/12/Perda-Kota-Semarang-Nomor-14-Tahun-2011.pdf.

24. It should be noted that, in 1864, the Dutch East Indies Railway Company erected Indonesia's first railway station on the same land. The station was later abandoned and passenger traffic moved to Stasiun Tawang, Semarang's largest train station to date. Long forgotten, today the flooded remnants of the train station only tickle the fancy of locomotive fans and hobby archaeologists.

25. In fact, the size of the retention area has been subject to contestation and reevaluation by public and state actors.

REFERENCES

Abram, Simone. 2014. "The Time It Takes: Temporalities of Planning." *Journal of the Royal Anthropological Institute* 20 (S1): 129–47.
Al Jazeera. 2015. "Pope: 'Catastrophic' Climate if Paris UN Talks Fail." November 27. https://www.aljazeera.com/news/2015/11/27/pope-catastrophic-climate-if-paris-un-talks-fail.
Anderson, Benedict. 1983. *Imagined Communities: Reflections on the Origin and Spread of Nationalism.* London: Verso.
Anand, Nikhil. 2017. "The Banality of Infrastructure." Social Science Research Council, "Just Environments," June 26. https://items.ssrc.org/just-environments/the-banality-of-infrastructure/.
Anand, Nikhil, Akhil Gupta, and Hannah Appel. 2018. *The Promise of Infrastructure.* Durham, N.C.: Duke University Press.
Asian Development Bank. 2012. *The Neighborhood Upgrading and Shelter Sector Project in Indonesia: Sharing Knowledge on Community-Driven Development.* Manila: Asian Development Bank.
Aspinall, Edward. 2013. "A Nation in Fragments: Patronage and Neoliberalism in Contemporary Indonesia." *Critical Asian Studies* 45 (1): 27–54.
Ballestero, Andrea. 2019. "The Underground as Infrastructure." In *Infrastructure, Environment, and Life in the Anthropocene: Experimental Futures,* edited by Kregg Hetherington, 17–44. Durham, N.C.: Duke University Press.
Barker, Joshua. 1998. "State of Fear: Controlling the Criminal Contagion in Suharto's New Order." *Indonesia,* no. 66 (October): 6–43.

Barker, Joshua. 2005. "Engineers and Political Dreams: Indonesia in the Satellite Age." *Current Anthropology* 46 (5): 703–27.
Barker, Joshua. 2007. "Vigilantes and the State." In *Identifying with Freedom: Indonesia after Suharto,* edited by Tony Day, 87–93. New York: Berghahn Books.
Barker, Joshua, Erik Harms, and Johan Lindquist. 2013. "Introduction to Special Issue: Figuring the Transforming City." *City & Society* 25 (2): 159–72.
Bauman, Zygmunt. (2003) 2011. *Wasted Lives: Modernity and Its Outcasts.* Cambridge: Polity.
Baxstrom, Richard. 2012. "Living on the Horizon of the Everlasting Present: Power, Planning and the Emergence of Baroque Forms of Life in Urban Malaysia." In *Southeast Asian Perspectives on Power,* edited by Liana Chua et al., 135–50. London: Routledge.
Beck, Ulrich. 1992. *Risk Society: Towards a New Modernity.* London: SAGE.
Benjamin, Walter. 1968. *Illuminations.* New York: Schocken Books.
Bhattacharyya, Debjani. 2018. *Empire and Ecology in the Bengal Delta: The Making of Calcutta.* Cambridge: Cambridge University Press.
Biehl, João Guilherme, and Peter Andrew Locke, eds. 2017. *Unfinished: The Anthropology of Becoming.* Durham, N.C.: Duke University Press.
Björkman, Lisa. 2015. *Pipe Politics, Contested Waters: Embedded Infrastructures of Millennial Mumbai.* Durham, N.C.: Duke University Press.
BKM Kemijen Mandiri. 2014. "Dokumen Rencana PLPBK Kelurahan Kemijen 2014–2015." Semarang.
Boomgaard, Peter. 2007. *A World of Water: Rain, Rivers and Seas in Southeast Asian Histories.* Leiden: Brill.
Bowen, John R. 1986. "On the Political Construction of Tradition: *Gotong Royong* in Indonesia." *Journal of Asian Studies* 45 (3): 545–61.
BPP Banger Sima. 2012. *Sima News Edisi Mai.* Kota Semarang.
Brett-Crowther, M. R. 1983. "Pondering on Polders." *Food Policy* 8 (2): 157–59.
Brommer, Bea, et al. 1995. *Semarang: Beeld van een stad.* Purmerend, Netherlands: Asia Maior.
Budiman, Manneke, Yuka D. N. Mangoenkoesoemo, P. Ayu Indah Wardhani, and Nila Ayu Utami. 2012. "New Enemy of the State: Youth in Post–New Order Indonesia." *Panorama: Insights into Asian and European Affairs* 1: 51–69.
Bunnell, Tim, Michelle Ann Miller, Nicholas A. Phelps, and John Taylor. 2013. "Urban Development in a Decentralized Indonesia: Two Success Stories?" *Pacific Affairs* 86 (4): 857–76.
Cazdyn, Eric. 2007. "Disaster, Crisis, Revolution." *South Atlantic Quarterly* 106 (4): 647–62.

Cazdyn, Eric. 2012. *The Already Dead: The New Time of Politics, Culture, and Illness.* Durham, N.C.: Duke University Press.

Chiotis, Theodoros, ed. and trans. 2015. *Futures: Poetry of the Greek Crisis.* London: Penned in the Margins.

Choy, Timothy. 2011. *Ecologies of Comparison: An Ethnography of Endangerment in Hong Kong.* Durham, N.C: Duke University Press.

Cobban, James L. 1974. "Uncontrolled Urban Settlement: The Kampong Question in Semarang (1905–1940)." *Bijdragen Tot de Taal-, Land- En Volkenkunde* 130 (4): 403–27.

Cobban, James L. 1988. "Kampungs and Conflict in Colonial Semarang." *Journal of Southeast Asian Studies* 19 (2): 266–91.

Colombijn, Freek. 2002. "Introduction: On the Road." *Bijdragen tot de taal-, land-en volkenkunde* 158 (4): 595–617.

Colven, Emma. 2020. "Subterranean Infrastructures in a Sinking City: The Politics of Visibility in Jakarta." *Critical Asian Studies* 52 (3): 311–31.

Cooper, Melinda E. 2008. *Life as Surplus: Biotechnology and Capitalism in the Neoliberal Era.* Seattle: University of Washington Press.

Coté, Joost. 2002. "Towards an Architecture of Association: H. F. Tillema, Semarang and the Construction of Colonial Modernity." In *The Indonesian Town Revisited,* edited by P. J. Nas, 319–47. Singapore: Institute of Southeast Asian Studies.

Coté, Joost. 2006. "Staging Modernity: The Semarang International Colonial Exhibition, 1914." *Review of Indonesian and Malaysian Affairs* 40 (1): 1–44.

Coté, Joost. 2014. "Thomas Karsten's Indonesia: Modernity and the End of Europe, 1914–1945." *Bijdragen tot de Taal-, Land- en Volkenkunde* 170 (1): 66–98.

Darsono, Suseno, and Mestika. 2011. "Peran masyarakat pada pengelolaan Polder Banger." Paper presented at Pertemuan Ilmiah Tahunan HATHI, Ambon, Maluku, October 28–30.

Das, Veena, and Deborah Poole, eds. 2004. *Anthropology in the Margins of the State.* Santa Fe, N.Mex.: School for Advanced Research Press.

Davis, Mike. 2007. *Planet of Slums.* London: Verso.

Dean, Mitchell. 1998. "Risk, Calculable and Incalculable." *Soziale Welt* 49 (1): 25–42.

De Boeck, Filip. 2005. "The Apocalyptic Interlude: Revealing Death in Kinshasa." *African Studies Review* 48 (2): 11–31.

De Boeck, Filip. 2014. *Kinshasa: Tales of the Invisible City.* Photographs by Marie-Françoise Plissart. Leuven: Leuven University Press.

de Certeau, Michel. 1984. *The Practice of Everyday Life.* Berkeley: University of California Press.

Degen, Monica Montserrat, and Gillian Rose. 2012. "The Sensory Experiencing of Urban Design: The Role of Walking and Perceptual Memory." *Urban Studies* 49 (15): 3271–87.
de L'Estoile, Benoît. 2014. "'Money Is Good, but a Friend Is Better': Uncertainty, Orientation to the Future, and 'the Economy.'" *Current Anthropology* 55 (S9): S62–S73.
Diamond, Jared. 2005. *Collapse: How Societies Choose to Fail or Succeed*. New York: Penguin.
Eickhoff, Martijn, Donny Danardono, Tjahjono Rahardjo, and Hotmauli Sidabalok. 2017. "The Memory Landscapes of '1965' in Semarang." *Journal of Genocide Research* 19 (4): 530–50.
Direktorat Tata Kota dan Tata Daerah. 1989. Direktorat Jenderal Cipta Karya, Departemen Pekerjaan Umum. "Studi profil kawasan laut dan udara." Jakarta.
Ferguson, James. 1999. *Expectations of Modernity: Myths and Meanings of Urban Life on the Zambian Copperbelt*. Berkeley: University of California Press.
Fortun, Kim. 2013. "Disaster: Integration." Member Voices, *Fieldsights*, April 2. https://culanth.org/fieldsights/disaster-integration.
Foucault, Michel. 2003. *Society Must Be Defended: Lectures at the Collège de France, 1975–1976*. New York: Macmillan.
Garr, Daniel. 1989. "Indonesia's Kampung Improvement Program: Policy Issues and Local Impacts for Secondary Cities." *Journal of Planning Education and Research* 9 (1): 79–83.
Gershon, Ilana. 2005. "Seeing like a System: Luhmann for Anthropologists." *Anthropological Theory* 5 (2): 99–116.
Gibbings, Sheri Lynn. 2021. *Shadow Play: Information Politics in Urban Indonesia*. Toronto: University of Toronto Press.
Giblett, Rodney J. 1996. *Postmodern Wetlands: Culture, History, Ecology*. Edinburgh: Edinburgh University Press.
Guinness, Patrick. 2009. *Kampung, Islam and State in Urban Java*. Singapore: NUS Press.
Gupta, Akhil. 2012. *Red Tape: Bureaucracy, Structural Violence, and Poverty in India*. Durham, N.C.: Duke University Press.
Guyer, Jane. 2007. "Prophecy and the Near Future: Thoughts on Macroeconomic, Evangelical, and Punctuated Time." *American Ethnologist* 34 (3): 409–21.
Hadiwinata, Bob S. 2003. *The Politics of NGOs in Indonesia: Developing Democracy and Managing a Movement*. London: Routledge Curzon.
Hadiz, Vedi R. 2004. "Decentralization and Democracy in Indonesia: A Critique of Neo-Institutionalist Perspectives." *Development and Change* 35 (4): 697–718.

Hadiz, Vedi R. 2006. "The Left and Indonesia's 1960s: The Politics of Remembering and Forgetting." *Inter-Asia Cultural Studies* 7 (4): 554–69.

Hansen, Thomas Blom. 2001. *Wages of Violence: Naming and Identity in Postcolonial Bombay.* Princeton, N.J.: Princeton University Press.

Harms, Erik. 2013. "Eviction Time in the New Saigon: Temporalities of Displacement in the Rubble of Development." *Cultural Anthropology* 28 (2): 344–68.

Harvey, David. 1989. "From Managerialism to Entrepreneurialism: The Transformation of Urban Governance in Late Capitalism." *Geografiska Annaler* 71 (1): 3–17.

Hendriks, Frank. 2001. *Polder Politics: The Re-invention of Consensus Democracy in the Netherlands.* New York: Routledge.

Heryanto, Ariel, and Nancy Lutz. 1988. "The Development of 'Development.'" *Indonesia*, no. 46 (October): 1–24.

Holgersson, Helena. 2016. "Challenging the Hegemonic Gaze on Foot: Walk-Alongs as a Useful Method." In *Walking in the European City: Quotidian Mobility and Urban Ethnography*, edited by Timothy Shortell and Evrick Brown, 207–25. London: Routledge.

Howe, Cymene, et al. 2016. "Paradoxical Infrastructures: Ruins, Retrofit, and Risk." *Science, Technology, & Human Values* 41 (3): 547–65.

Ingleson, John. 2001. "The Legacy of Colonial Labour Unions in Indonesia." *Australian Journal of Politics and History* 47 (1): 85–100.

ISET (Institute for Social and Environmental Transition). 2011. "Semarang Climate Change Resilience Strategy, Indonesia." June.

Jackson, Michael, and Albert Piette. 2015. *What Is Existential Anthropology?* New York: Berghahn Books.

Jackson, Steven J. 2014. "Rethinking Repair." In *Media Technologies: Essays on Communication, Materiality, and Society*, edited by Tarleton Gillespie, Pablo J. Boczkowski, and Kirsten A. Foot, 221–39. Cambridge, Mass.: MIT Press.

Katz, Jack. 2001. "From How to Why: On Luminous Description and Causal Inference in Ethnography (Part I)." *Ethnography* 2 (4): 443–73.

Kementerian Pekerjaan Umum dan Perumahan Rakyat. 2019. "Kementerian PUPR Targetkan Infrastruktur Pengendali Banjir Rob Kota Semarang Rampung Pertengahan 2019." https://pu.go.id/berita/kementerian-pupr-targetkan-infrastruktur-pengendali-banjir-rob-kota-semarang-rampung-pertengahan-2019%C2%A0.

Khosravi, Shahram. 2019. "What Do We See if We Look at the Border from the Other Side?" *Social Anthropology* 27 (3): 409–24.

Kinanti, Melody Kristianai. 2013. "Kajian Bentuk Lingkungan Permukiman Berdasarkan *Sense of Community* di Kelurahan Dadapsari Semarang." Thesis, Diponegoro University, Semarang.

Klein, Naomi. 2014. *This Changes Everything: Capitalism vs. the Climate.* New York: Simon & Schuster.
Kleinman, Julie. 2014. "Adventures in Infrastructure: Making an African Hub in Paris." *City & Society* 26 (3): 286–307.
Kooy, Michelle, and Karen Bakker. 2008. "Technologies of Government: Constituting Subjectivities, Spaces, and Infrastructures in Colonial and Contemporary Jakarta." *International Journal of Urban and Regional Research* 32 (2): 375–91.
Kurniawan, Eka. 2015. *Beauty Is a Wound.* New York: New Directions.
Kusno, Abidin. 2000. *Behind the Postcolonial: Architecture, Urban Space, and Political Cultures in Indonesia.* New York: Routledge.
Kusno, Abidin. 2012. "Housing the Margin: Perumahan Rakyat and the Future Urban Form of Jakarta." *Indonesia,* no. 94 (October): 23–56.
Kusno, Abidin. 2018. "Where Will the Water Go?" *Indonesia,* no. 105 (April): 19–51.
Kusuma, Y., N. Venketasubramanian, L. S. Kiemas, and J. Misbach. 2009. "Burden of Stroke in Indonesia." *International Journal of Stroke* 4 (5): 379–380.
Lainez, Nicolas. 2019. "Treading Water: Street Sex Workers Negotiating Frantic Presents and Speculative Futures in the Mekong Delta, Vietnam." *Time & Society* 28 (2): 804–27.
Lassa, Jonatan A. 2012. "Semarang Joins the 'Sinking Cities Network.'" *Jakarta Post,* October 6.
Ley, Lukas. 2018. "On the Margins of the Hydrosocial: Quasi-Events along a Stagnant River." *Geoforum,* April. https://doi.org/10.1016/j.geoforum.2018.03.010.
Ley, Lukas. 2020. "Figuring (Out) the Sinking City: Tidal Floods and Urban Subsidence in Semarang, Indonesia." In *Disastrous Times: Beyond Environmental Crisis in Urbanizing Asia,* edited by Tyson Vaughan and Eli Elinoff, 46–64. Philadelphia: University of Pennsylvania Press.
Ley, Lukas. 2021. "The Global Swamp. Or, the Amphibious as a Figure of Heterotopia." In *Delta Life: Exploring Dynamic Environments where Rivers Meet the Sea,* edited by Franz Krause and Mark Harris. London: Berghahn Books.
Li, Tania Murray. 2005. "Beyond 'the State' and Failed Schemes." *American Anthropologist* 107 (3): 383–94.
Li, Tania Murray. 2007. *The Will to Improve: Governmentality, Development, and the Practice of Politics.* Durham, N.C.: Duke University Press.
Li, Tania Murray. 2015. "Governing Rural Indonesia: Convergence on the Project System." *Critical Policy Studies* 10 (1): 79–94.

Logsdon, Martha Gay. 1974. "Neighborhood Organization in Jakarta." *Indonesia*, no. 18 (October): 53–70.

Lucas, Anton. 2000. *The Dog Is Dead, so Throw It in the River: Environmental Politics and Water Pollution in Indonesia.* With Ariel Djati. Papers on Southeast Asia 51. Clayton, Vict., Australia: Monash Asia Institute.

Marfai, Muh Aris, and Lorenz King. 2008. "Tidal Inundation Mapping under Enhanced Land Subsidence in Semarang, Central Java Indonesia." *Natural Hazards* 44: 93–109.

Marfai, Muh Aris, Lorenz King, Junun Sartohadi, Sudrajat Sudrajat, Sri Rahayu Budiani, and Fajar Yulianto. 2008. "The Impact of Tidal Flooding on a Coastal Community in Semarang, Indonesia." *The Environmentalist* 28 (3): 237–48.

Masco, Joseph. 2010. "Bad Weather: On Planetary Crisis." *Social Studies of Science* 40 (1): 7–40.

Masco, Joseph. 2012. "The End of Ends." *Anthropological Quarterly* 85 (4): 1107–24.

Masco, Joseph. 2017. "The Crisis in Crisis." *Current Anthropology* 58 (S15): S65–S76.

Mingu. 2016. "Perumahan Kumuh Kemijen di Semarang Ini Disulap Jadi Beda, Hasilnya Mengejutkan." *Tribun Jateng*, July 31. https://jateng.tribunnews.com/2016/07/31/perumahan-kumuh-kemijen-di-semarang-ini-disulap-jadi-beda-hasilnya-mengejutkan.

Mitchell, Timothy. 2002. *Rule of Experts: Egypt, Techno-politics, Modernity.* Berkeley: University of California Press.

Mondeel, Herman, and Hermono S. Budinetro. 2010. *The Banger Polder in Semarang.* Solo, Central Java, Indonesia: Center for River Basin Organizations and Management.

Morita, Atsuro. 2017. "Multispecies Infrastructure: Infrastructural Inversion and Involutionary Entanglements in the Chao Phraya Delta, Thailand." *Ethnos* 82 (4): 738–57.

Mrázek, Rudolf. 2002. *Engineers of Happy Land: Technology and Nationalism in a Colony.* Princeton, N.J.: Princeton University Press.

Mrázek, Rudolf. 2016. Review of *Beauty Is a Wound: A Novel*, by Eka Kurniawan. *Indonesia*, no. 101 (April): 145–50.

Muhammad, Djawahir. 2011. *Membela Semarang!* Semarang: Pustaka Semarang 16.

Murti, Yoshi Fajar Kresno. 2015. "'Babad Kampung': Celebrating History and Neighborhood Identity in Yogyakarta." In *Performing Contemporary Indonesia: Celebrating Identity, Constructing Community*, edited by Barbara Hatley and Brett Hough, 45–67. Leiden: Brill.

Narotzky, Susana, and Niko Besnier. 2014. "Crisis, Value, and Hope: Rethinking the Economy." *Current Anthropology* 55 (S9): S4–S16.
Nas, Peter J. M. 2003. *The Indonesian Town Revisited.* Münster: LIT Verlag.
Nestor, James. 2014. *Deep: Freediving, Renegade Science, and What the Ocean Tells Us about Ourselves.* New York: Houghton Mifflin Harcourt.
Newberry, Jan. 2006. *Back Door Java: State Formation and the Domestic in Working Class Java.* Toronto: University of Toronto Press.
Newberry, Jan. 2008. "Double Spaced: Abstract Labour in Urban Kampung." *Anthropologica* 50 (2): 241–53.
Newberry, Jan. 2018. "A Kampung Corner: Infrastructure, Affect, Informality." *Indonesia,* no. 105 (April): 191–206.
Oppenheimer, Joshua, dir. 2014. *The Look of Silence.* Documentary. 103 min.
Padawangi, Rita, and Mike Douglass. 2015. "Water, Water Everywhere: Toward Participatory Solutions to Chronic Urban Flooding in Jakarta." *Pacific Affairs* 88 (3): 517–50.
Peck, Jamie. 2014. "Entrepreneurial Urbanism: Between Uncommon Sense and Dull Compulsion." *Geografiska Annaler: Series B, Human Geography* 96 (4): 396–401.
Peters, Robin. 2012. *Factors That Contribute to Effective Dutch-Funded International Water Projects: A Case Study: Banger Pilot Polder Project in Semarang, Indonesia.* Enschede, the Netherlands: Universiteit Twente.
Povinelli, Elizabeth. 2011. *Economies of Abandonment: Social Belonging and Endurance in Late Liberalism.* Durham, N.C.: Duke University Press.
Povinelli, Elizabeth, Julieta Aranda, Brian Kuan Wood, and Anton Vidokle. 2014. "Editorial—'Quasi-Events.'" *E-Flux,* no. 58. http://www.e-flux.com/journal/58/61143/editorial-quasi-events/.
Pratiwo. 2004. "The City Planning of Semarang 1900–1970." Paper presented at the First International Urban Conference, Surabaya.
Prudham, Scott. 2009. "Pimping Climate Change: Richard Branson, Global Warming, and the Performance of Green Capitalism." *Environment and Planning A* 41 (7): 1594–1613.
Rabinow, Paul. 1995. *French Modern: Norms and Forms of the Social Environment.* Chicago: University of Chicago Press.
Rademacher, Anne. 2011. *Reigning the River: Urban Ecologies and Political Transformation in Kathmandu.* Durham, N.C.: Duke University Press.
Reid, Anthony. 1993. *Southeast Asia in the Age of Commerce, 1450–1680: Expansion and Crisis.* New Haven, Conn.: Yale University Press.
Richter, Simon. 2020. "Translation of *Polder*: Water Management in The Netherlands and Indonesia." In *The Oxford Handbook of Translation and*

Social Practices, edited by Meng Ji and Sara Laviosa, n.p. Oxford: Oxford University Press. https://www.oxfordhandbooks.com.

Roberts, John. 2005. "Fuel Price Rise in Indonesia Triggers Protests." International Committee of the Fourth International, *World Socialist Web Site,* April 1. https://www.wsws.org/en/articles/2005/04/indo-a01.html.

Rohmah, Ainur. 2013. "Semarang to Build Polder to Tackle Floods." *Jakarta Post,* July 26.

Roitman, Janet. 2013. *Anti-crisis.* Durham, N.C.: Duke University Press.

Roosmalen, Pauline K. M. van. 2008. "For Kota and Kampong: The Emergence of Town Planning as a Discipline." PhD diss., Institute for the History of Art, Architecture and Urbanism, TU Delft, Leiden.

Scott, James C. 1998. *Seeing like a State: How Certain Schemes to Improve the Human Condition Have Failed.* Yale Agrarian Studies. New Haven, Conn.: Yale University Press.

Semarang Municipality. 2016. "SEMARANG TANGGUH: Bergerak Bersama Menuju Semarang Tangguh." Balai Kota Semarang.

Sharma, Sarah. 2014. *In the Meantime: Temporality and Cultural Politics.* Durham, N.C.: Duke University Press.

Siegel, James T. 1993. *Solo in the New Order: Language and Hierarchy in an Indonesian City.* Princeton, N.J.: Princeton University Press.

Siegel, James T. 1997. *Fetish, Recognition, Revolution.* Princeton, N.J.: Princeton University Press.

Siegel, James T. 1998. *A New Criminal Type in Jakarta: Counter-revolution Today.* Durham, N.C.: Duke University Press.

Silver, Christopher. 2011. *Planning the Megacity: Jakarta in the Twentieth Century.* London: Routledge.

Simone, AbdouMaliq. 2010. *City Life from Jakarta to Dakar: Movements at the Crossroads.* London: Routledge.

Simone, AbdouMaliq. 2020. "To Extend: Temporariness in a World of Itineraries." *Urban Studies* 57 (6): 1127–42.

Star, Susan Leigh. 1999. "The Ethnography of Infrastructure." *American Behavioral Scientist* 43 (3): 377–91.

Stern, Lesley. 2017. "A Garden or a Grave: The Canyonic Landscape of the Tijuana–San Diego Region." In *Arts of Living on a Damaged Planet: Ghosts and Monsters of the Anthropocene,* edited by Anna Tsing, Heather Swanson, Elaine Gan, and Nils Bubandt, 17–29. Minneapolis: University of Minnesota Press.

Stoler, Ann Laura. 1995. *Race and the Education of Desire: Foucault's History of Sexuality and the Colonial Order of Things.* Durham, N.C.: Duke University Press.

Suara Merdeka. 2016. "Jokowi Pantau Rob di Pantura Semarang." June 20.

Sullivan, John. 1986. "Kampung and State: The Role of Government in the Development of Urban Community in Yogyakarta." *Indonesia*, no. 41 (April): 63–88.

Swyngedouw, Erik. 2010. "Apocalypse Forever?" *Theory, Culture & Society* 27 (2–3): 213–32.

Taussig, Michael T. 2015. *The Corn Wolf.* Chicago: University of Chicago Press.

Thompson, E. P. 1975. *Whigs and Hunters: The Origin of the Black Act.* New York: Pantheon Books.

Tillema, H. F. 1911. *Rioliana.* Semarang: n.p.

Tillema, H. F. 1913. *Van wonen en bewonen, van bouwen, huis en erf.* Tjandi-Samarang: n.p.

Trumbull, Raissa DeSmet. 2013. "A Liquid World: Figuring Coloniality in the Indies." PhD diss., University of California, Santa Cruz.

Tsing, Anna Lowenhaupt. 1993. *In the Realm of the Diamond Queen: Marginality in an Out-of-the-Way Place.* Princeton, N.J.: Princeton University Press.

Tsing, Anna Lowenhaupt. 2005. *Friction: An Ethnography of Global Connection.* Princeton, N.J.: Princeton University Press.

Tsing, Anna Lowenhaupt. 2015. *The Mushroom at the End of the World: On the Possibility of Life in Capitalist Ruins.* Princeton, N.J.: Princeton University Press.

UNDP Indonesia. 2017. "Good Practices: Integrating the SDGS into Development Planning." https://un.info.np/Net/NeoDocs/View/8009.

Van der Pal, Eelco. 2012. "Enhancing the Applicability of the Polder Concept." MSc thesis, Delft University of Technology.

Van Marwijk Advies. 2014. "From Aid to Trade in the Lowlands: Opportunities for the Dutch Water Sector to Cooperate with the Productive Sector in the Lowlands in Indonesia." Report for the Ministry of Infrastructure and the Environment / Deltateam Indonesia and the Netherlands Water Partnership.

Venkatesh, Sudhir. 2015. "On Urban Ethnography." *City & Community* 14 (4): 347.

Versnel, Hans, and Freek Colombijn. 2014. "Rückert and Hoesni Thamrin: Bureaucrat and Politician in Colonial Kampong Improvement." In *Cars, Conduits, and Kampongs: The Modernization of the Indonesian City, 1920–1960,* edited by Freek Colombijn and Joost Coté, 121–51. Leiden: Brill.

Vickers, Adrian. 2005. *A History of Modern Indonesia.* Cambridge: Cambridge University Press.

Voorst, Roanne van. 2014. "The Right to Aid: Perceptions and Practices of Justice in a Flood-Hazard Context in Jakarta, Indonesia." *Asia Pacific Journal of Anthropology* 15 (4): 339–56.

Wertheim, Willem Frederik, and The Siauw Giap. 1962. "Social Change in Java, 1900–1930." *Pacific Affairs* 35 (3): 223–47.

Wicitra, Joan. 2014. "Commitment to the Agents of Change." *Jakarta Post*, August 26.

Wilk, Richard. 2007. "It's about Time: A Commentary on Guyer." *American Ethnologist* 34 (3): 440–43.

Witteveen+Bos. 2014. "Site Visit Report Semarang." Unpublished document in author's possession.

Woldendorp, Jaap, and Hans Keman. 2007. "The Polder Model Reviewed: Dutch Corporatism 1965–2000." *Economic and Industrial Democracy* 28 (3): 317–47.

Yapp, Lauren. 2020. "To Help or Make Chaos? An Ethnography of Dutch Expertise in Postcolonial Indonesia." In *Heritage as Aid and Diplomacy in Asia*, edited by Philippe Peycam, Shu-Li Wang, and Hsin-Huang Michael Hsiao, 143–66. Singapore: ISEAS.

INDEX

Abram, Simone, 161, 181
activism, 27, 29, 126, 127, 161, 197; Komayu (resident organization), 136, 140–42, 143; network, 132
Adin, 39–40, 144, 175, 178, 191; ethnographic methods, 26; eviction, 170, 184, 186, 188; festivals, 76, 141; Komayu (resident organization), 75, 142, 184; mangroves, 41, 55, 74; pumping house (polder), 177, 179; SIMA (water authority), 160, 183, 185; subsystem survey, 173, 174. *See also* Indah; Sumarmo
agency, 86, 195
airport, international, 6
Aliansi (Civil Alliance Against Flooding), 136, 143, 144, 145
alliances, 29, 126, 130, 136, 142, 143
Anand, Nikhil, 37

anthropology, 18–19, 20, 25, 161, 194; "existential anthropology," 110. *See also* ethnographic methods
antiflooding: labor, 22; measures, 30, 77, 100, 143, 151, 160; policy, 123. *See also* maintenance; polder (pilot project); repair, constant
apocalyptic potential, 5, 78, 86, 87, 88, 163; sea level rise, 13
aquaculture, 43, 45, 53, 180
architectures of time, 5, 86, 109
archival research methods, 19, 30, 36, 43, 206n4
Arief, 55, 96–101, 104–5, 184; grants, 149; labor, "community," 22, 134; meetings, 66; *pompanisasi* (community-run water pump group), 2, 134; *rob,* effects of, 97, 102; skepticism, 192; *warung* (canteen), 102–3, 110, 135

227

Index

Ariel, 1, 2, 5, 7, 22; grants, 149; *rob*, effects of, 4; *warung* (canteen), 96, 102, 192
Aspinall, Edward, 128, 129, 138, 209n4

Badan Nasional Penanggulangan Bencana (National Disaster Management Agency), 160
balancing act, 122–26
Ballestero, Andrea, 196
Banger River description, 41–42, 72. *See also* normalization (*normalisasi*), concept of; pollution
Bappeda (Semarang's Planning Agency), 15, 40, 60, 172, 191, 193; civil participation, 175; future, imagined, 80; problem zones, 62; PSDA (Semarang's water agency), 143; water tax, 81
Barker, Joshua, 69, 70, 71, 126, 148, 174
"baroque mode," 87, 109, 194
Baxstrom, Richard, 87, 93, 103, 109, 159
belongings, personal, 3, 7, 35
Benjamin, Walter, 163
Besnier, Niko, 88
Björkman, Lisa, 60, 135
BKM (Badan Keswadayaan Masyarakat), 23, 65, 147, 209n2
"black lists," 65, 173
Boomgaard, Peter, 195
"borrowed time," 6, 12, 106, 120, 158, 193; ethnographic methods, 196; funding, 150; "quasi-events" (discussion), 194; repair, constant, 7, 9
bottom-up development. *See* grassroots politics
Bowen, John, 20, 21, 205n3
bric-a-brac, 90–95, 101–4
bridges, 71, 132, 133, 191
Budiman, Manneke, 69, 70

canteen. *See warung* (canteen)
capitalism, 10, 51, 86, 163–64, 181; decentralization, 129; marginalization, 53, 54; neoliberal values, 130, 166, 194; urbanization, 154, 195
capitalist reproduction, 10, 163
care, communal, 19, 53, 58, 163; festivals, 77, 78, 148
catch-22, 88, 120
causes of flooding, 106, 119, 132, 142, 173, 210n8; mangroves, 85; rainfall, 38, 92; river walls, 133; sea level rise, 78, 131. *See also* groundwater extraction
Cazdyn, Eric, 9–11, 163, 188. *See also* chronic present; "meantime"
change, institutional, 158, 163, 189
change, social, 4, 119, 131, 155
Chinese, 17, 28, 43, 209n4, 210n7
choice, 12, 26, 87, 185
chronic present, 3, 5, 9, 10–12, 30, 31
citizen responsibility. *See* responsibility, resident
city council, Semarang's, 4, 17, 46, 48, 50, 82
City Hall, 55, 101, 160, 172, 186, 192
Civil Alliance Against Flooding (Aliansi Masyarakat Terhadap Rob dan Banjir), 136, 143, 144, 145
civil participation, 19, 96, 122, 149, 175, 176. *See also* festivals; meetings

cleaning, 2, 19, 58, 71, 77
climate change, 29, 31, 38, 157, 194, 196; festivals, 76, 142, 145, 170. *See also* global warming; Intergovernmental Panel on Climate Change (IPCC)
clogging, 4, 25, 34, 59, 95, 192
coastal neighborhoods, 30, 31, 39, 43, 72, 109; fishponds, 35, 192; seeping riverbanks, 4
Cobban, James, 46, 47, 54, 63, 99, 207n4
Colombijn, Freek, 17, 44, 50
colonial government, 36, 44, 47, 49, 50, 61
Colven, Emma, 7, 126
"community" labor (*kerja bakti*). *See* labor, "community"
community meetings, 67, 137
community-run, 2, 9, 12, 186. *See also pompanisasi* (community-run water pump group)
compensation, 74, 159, 170, 174, 181, 184
coordination, 11, 21, 23, 70, 122, 145; repair, constant, 11; Wahyu, 138, 140, 146; Imam Wahyudi (water expert), 162, 172. *See also pompanisasi* (community-run water pump group)
corruption, 23, 75, 142, 179, 180, 181; KPK (Corruption Eradication Commission), 178
Coté, Joost, 17, 44, 46, 47, 48, 49
criminality, 28, 62, 67, 68, 70, 71; killings, mass, 64, 65; Komayu (resident organization), 141; normalization (*normalisasi*), concept of, 69, 74, 75; Wahyu, 137. *See also* violence

crisis, action from, 29, 153, 161, 162, 182, 189. *See also* Cazdyn, Eric; De Boeck, Filip
crisis assessment, 13, 14, 164, 169, 188
crisis imaginaries, 13, 30, 161, 164, 188
crisis narratives, 154, 162, 163, 169

daily rhythms, 27, 38, 106, 176, 183, 192; chronic present, 3, 11; criminality, 68; ethnographic methods, 25, 26, 60, 61, 90, 208–9n4; exhaustion, 2, 96, 110; knowledge (deep), 173; *rob*, effects of, 39; stacking, 91. *See also* cleaning
dams, 28, 79, 180, 185, 191, 193; northern polder dam, 169
danger, 38, 70, 137, 151, 163, 167; criminality, 68; health, 9, 20, 53, 62, 195; killings, mass, 64; resistance, 75; social control, 69; stigmatization, 63, 195; violence, 50, 51, 61, 62; waterborne diseases, 3, 50. *See also* endangered
"dark," 63, 66, 82, 136, 141, 195
Darsono, Suseno (SIMA chairman), 79, 172, 175
De Boeck, Filip, 87–88
debt, 7, 8, 105–6
decentralization, 47, 122, 149, 165, 209n3, 210n6; capitalism, 129; governance, participatory, 125, 137; "Musrenbang" (participatory development scheme), 126; New Order, 131; transparency, 124, 139
delta, 43, 44, 70, 71, 82, 121; apocalyptic potential, 78; description,

19, 46, 86; "hard," 110; maps, 81; marginalization, 195; sea level rise, 158
democratic reform, 22, 150, 154, 164, 165, 193; "multi-stakeholder" approach, 152
Deni, 90–95, 98, 110, 184
development plans, 17, 90, 93, 109, 142–45; urban redevelopment, 5, 16; World Bank, 129
diesel, 2, 11, 29, 98, 134
differentiation, 28, 37, 54, 60, 89, 106; racial differentiation, 28, 54
"dirty," 2, 4, 7, 49, 71
diseases, waterborne, 2, 3, 46, 50
dispossession, 20, 193
domestic violence, 137
donations, private, 14, 23, 24
drinking water, 25, 49, 195, 211n6
Dutch government, 37, 46, 164, 166–67, 172; and Johan Helmer (polder project leader), 166, 167; polder board, 168
duty, 21, 47, 110, 124

East Flood Canal, 41, 42, 44, 56, 74, 191; overflow, 35, 186
ecological deterioration, 28, 37, 82, 110, 153, 169; crisis imaginaries, 161; mangroves, 55, 85; pollution, 59; "rent to nature," 150; urbanization, 41
economic stagnation, 12, 53, 92, 104, 153
EcoShape, 85, 208n2
Edo, 104
Eko (landlord), 57, 59, 74
embankments (*talut*), 68, 95, 121, 133–34, 146; criminality, 71; *pompanisasi* (community-run water pump group), 134, 136; stacking, 91
endangered, 29, 143, 145, 162, 196
Eny, 92, 93, 95
epidemics, 20, 47, 54, 63, 206–7n6
ethnicity, 15, 17, 53, 135, 205n4
ethnographic methods, 12, 24–26, 194, 196; photo-ethnographic methods, 90, 109; urban research methods, 174, 208–9n4
eviction, 7, 28, 75, 87, 180, 186; compensation, 159, 170, 181, 184; "normalization of violence," 164, 188; plans, 5, 103; retention basin (polder), 139, 157, 181, 184
exhaustion, 2, 12, 18, 89, 119, 185; Arief, 96, 100; Imam, 110
existence, mode of, 13, 15
"existential anthropology," 110
experts (water), 4, 79, 151, 152, 153, 174. *See also* Helmer, Johan (polder project leader); Mondeel, Herman (water engineer); Wahyudi, Imam (water expert)
extraction, groundwater. *See* groundwater extraction

faskel (PNPM facilitator), 23, 209n2
Ferguson, James, 60, 154
Festival Rakyat (People's Festival), 147
festivals, 27, 75–78, 141, 142, 145; Climate Change Festival, 170; Festival Kali Banger, 148; Festival Rakyat (People's Festival), 147; funding, 24; National Holiday (Hari Kemerdekaan), 26; Penta K Labs, 151

fishponds, 35, 44, 52, 181, 184, 192; ecological deterioration, 41; eviction, 139
Foucault, Michel, 51
funding, 75, 77, 121, 134, 150, 179; civil participation, 175; competition, 125; cycles, 11; Dutch government, 167; eligibility, 59; Mercy Corps (NGO), 24, 76; "Musrenbang" (participatory development scheme), 30, 138; subdistrict government (*kelurahan*), 122. *See also* grants
future, imagined, 80, 129, 150, 158, 163; crisis imaginaries, 161, 164

global capitalism, 164, 166
global warming, 78, 158, 161, 168, 195. *See also* sea level rise
gotong-royong (mutual assistance), 20, 21, 177, 205n3
governance, participatory, 29, 125, 137, 164, 174, 188
governance, water. *See* water governance
government grants, 18, 23, 98, 192; state development grants, 29, 126, 138
grants, 14, 132, 146, 147, 149, 177; international granting agencies, 130; NGOs (nongovernmental organizations), 140; PNPM Mandiri (National Program for Community Empowerment), 135; state development grants, 29, 126, 138
grassroots politics, 127, 129, 137, 139–40, 145; alliances, 29, 126, 130, 136, 142, 143; Komayu (resident organization), 136, 145, 150
groundwater extraction, 6, 7, 72, 73, 158, 196
Guinness, Patrick, 22, 23, 61
Gupta, Akhil, 37, 110
Guyer, Jane, 181

"hard," 86, 109, 110, 120, 123
hardship, 29, 53, 86, 87, 120, 208n4
health, 9, 18, 48, 53, 62, 68; pollution, 20, 195. *See also* strokes; waterborne diseases
Helmer, Johan (polder project leader), 158, 165, 169, 171, 178, 187; and the Dutch government, 166, 167
Heryanto, Ariel, 149
hierarchies, power, 8, 126, 173, 180, 182
house adaptations, 56, 107–8, 109, 113–18; debt, 105–6; pride, 104; raising (*peninggian*), 8, 40, 111, 114, 116, 118
Hysteria (art collective), 151

identity, Javanese, 17, 172
ideologies, dominant, 6, 11, 125, 126, 128, 155
imagined future. *See* future, imagined
Imam, 102–3, 105–8, 110
income, sources of, 44, 46, 52, 110, 138, 148; and debt, 105; Festival Rakyat (People's Festival), 147; house adaptations, 107; land, access to, 28, 53; raising (*peninggian*), 187
Indah, 55–56, 172, 192
indigenous population, 37, 44, 48, 49, 50, 51; dispossession, 193;

land, access to, 53; relocation, 46, 54; stigmatization, 47, 63; urbanization, 43
Indonesian Ministry of Public Works, 30, 165, 171, 176
industrialization, 27, 72, 74. *See also* urbanization
industrial pollution, 24, 42, 72, 207n1
inequality, 5, 14, 132, 154. *See also* differentiation
infrastructural improvement projects, 11, 86, 93, 120, 122
inspection, 47, 63, 110, 171, 177; mayoral visits, 96, 100; roads, 55, 68, 70, 73
institutional change, 158, 163, 189
Intergovernmental Panel on Climate Change (IPCC), 79
interlocutors, 25, 31, 35, 37, 59, 62; "black lists," 65; catch-22, 120; criminality, 68; religion, 24; subdistrict government (*kelurahan*), 128
international development agencies, 22, 138
international granting agencies, 130
interview methods, 104, 144, 212n10
Islam, 24, 55
iuran (community tax), 98, 134, 177

Jackson, Michael, 110
Jackson, Steven, 7, 9, 87, 88, 95, 101
jalan santai event, 1, 4, 18
Javanese cities, 46, 51, 92

Kampung Improvement Program (KIP), 71, 207–8n5

kampungs, formation of, 18–21, 36, 42–43, 46
Kemijen subdistrict government. *See* subdistrict government (*kelurahan*)
killings, mass, 16, 17, 64–65
"killing" the river, 28, 78, 79, 82
knowledge (deep), 1, 41, 132, 135, 173, 182; ethnographic methods, 12; *pompanisasi* (community-run water pump group), 136; SIMA (water authority), 172
Komayu (resident organization), 75, 140–43, 150, 184; Aliansi (Civil Alliance Against Flooding), 136, 143, 145
KPK (Corruption Eradication Commission), 178
Kraft van Ermel, Roy (Dutch project worker), 156, 211n3
Kusno, Abidin, 7, 35, 36, 70, 81, 99
Kusuma, Yohanna, 51

labor, "community," 22, 98, 99, 110, 119, 134; exhaustion, 2; funding, 135, 207–8n5; *gotong-royong* (mutual assistance), 21, 177, 205n3; repair, constant, 95
land, access to, 20, 28, 53, 154
land, use of, 16, 44, 73, 108
land ownership, 46, 52, 53, 73, 74
Li, Tania Murray, 11, 13, 127, 129–31, 146
littoral, 19, 27, 78, 79, 81, 168
Logsdon, Martha Gay, 21
long-term thinking, 12, 98, 112, 156, 181, 188; architectures of time, 86; development plans, 5; global warming, 168; repair, constant, 89

LPMK (Neighborhood Community Empowerment Board), 58, 173
lurah (subdistrict head), 100–101, 122, 123, 124, 147; mayoral visits, 77–78, 96, 100; meetings, 58, 148
Lutz, Nancy, 149

maintenance, 7, 11, 80, 101, 159, 172; activities, 2, 31, 58, 124, 146; care, communal, 77; civil participation, 122, 175, 176; labor, "community," 22, 48, 98; SIMA (water authority), 173, 175; subdistrict government (*kelurahan*), 22; taxation, 98, 134. *See also* house adaptations; repair, constant
mangroves, 41, 55, 74, 85, 119
maps, 36, 43, 44, 45, 46, 48; "baroque mode," 87; delta, 81; littoral, 27; mortality rates, 50
marginalization, 19, 20, 92, 180, 207–8n5; and capitalism, 53–54, 195; political, 83, 99, 196; violence, 61, 65. *See also* stigmatization
markets, 15, 16, 65, 103
marshland, 27, 43, 53, 54, 62, 74; airport, international, 6
Masco, Joseph, 161, 162, 163, 188
mayoral visits, 58, 77–78, 96, 100, 151. *See also* Prihadi, Hendrar (Semarang mayor)
"meantime," 87, 88, 181, 187, 194, 196; and raising (*peninggian*), 30; repair, constant, 5, 14, 101, 120
meetings, 57–58, 59, 66, 81, 148; alliances, 142; community meetings, 27, 67, 137; "dark," 141; neighborhood meetings, 66, 77, 96, 99–100, 147; pilot project (polder), 160; provincial level, 85; public meetings, 123, 169; SIMA (water authority), 172, 175, 177, 183–84, 186. *See also* Aliansi (Civil Alliance Against Flooding); *sosialisasi* (community events); *warung* (canteen)
Mercy Corps (NGO), 24, 76, 142, 160
Mila, 33–35, 38–40
mobilization, 15, 21, 69, 89, 194, 195; "rent to nature," 8, 121
mold, 4, 33, 92, 118
Mondeel, Herman (water engineer), 156, 157, 167, 181
monsoon, 19, 38, 74, 106, 157
Morita, Atsuro, 19
mortality rates, 43, 50, 196, 207n7
mosques, 24
Mrázek, Rudolf, 44, 48, 49, 63, 163
Muhammad, Djawahir, 86
"multi-stakeholder" approach, 138, 152
municipality of Semarang, 79, 121, 141, 168, 171, 187; land ownership, 52; polder technology, 155; protest, 169. *See also* Bappeda (Semarang's Planning Agency); "Musrenbang" (participatory development scheme); PSDA (Semarang's water agency)
Murti, Yoshi Fajar Kresno, 127
"Musrenbang" (participatory development scheme), 29, 126, 137, 138, 139, 150
mutual assistance (*gotong-royong*), 20, 21, 177, 205n3

Narotzky, Susana, 88
National Disaster Management Agency (Badan Nasional Penanggulangan Bencana), 160
National Holiday (Hari Kemerdekaan), 26
National Program for Community Empowerment (Program Nasional Pemberdayaan Masyarakat Mandiri), 23, 135, 142, 146
neighborhood associations, 96, 142, 146
neighborhood meetings, 66, 77, 96, 99–100, 147
neoliberal values, 10, 126, 129, 176, 196; capitalism, 130, 166, 194
Nestor, James, 25–26
Newberry, Jan, 21, 61, 90, 99, 177
New Order, 65, 81, 127, 128, 131; and grassroots politics, 129–30, 136; Suharto (Indonesian president 1967–98), 28, 52, 64
NGOs (nongovernmental organizations), 85, 130, 136, 138, 140, 143; Mercy Corps (NGO), 24, 76, 142, 160; Pattiro (NGO), 123, 144, 145; Perdikan (local NGO), 76
normalization (*normalisasi*), concept of, 28, 69, 70, 75, 195, 196
"normalization of violence," 110, 163, 164, 188
northern polder dam, 169

overflow, 3, 5, 60, 92, 125, 192; East Flood Canal, 35, 186; sewage, 2

Pak Rianto, 51–53, 184
Pak Rozi, 52

participant observation, 25, 61, 90, 110, 123, 177
participatory development agenda, 22, 132, 155. *See also* "Musrenbang" (participatory development scheme)
participatory governance. *See* governance, participatory
Pattiro (NGO), 123, 144, 145
pembangunan (development), 30, 126, 137, 145; meaning, 8
peninggian. *See* raising (*peninggian*)
People's Festival. *See* Festival Rakyat (People's Festival)
Perdikan (local NGO), 76
Perserikatan Kommunist di India (PKI), 17, 163
Pertamina (state-owned company), 24, 72, 90
Petrus massacres, 64–65
photo-ethnographic methods, 90, 109
Piette, Albert, 110
planning agency. *See* Bappeda (Semarang's Planning Agency)
PNPM Mandiri (National Program for Community Empowerment), 23, 135, 142, 146
polder (pilot project), 4, 80, 159, 160, 165, 177; crisis imaginaries, 161; EcoShape, 85; governance, participatory, 164, 188
polder board, 156, 160, 171–75, 176, 192; Dutch government, 168; members, 170, 174, 177, 178, 186; water tax, 81, 159. *See also* SIMA (water authority)
polder technology, 14, 155, 156, 158, 159, 164
political marginalization, 83, 99, 196

political reforms, 22, 30, 31, 126, 150, 173
political transition, 125, 127, 129
pollution, 19, 60, 69, 86, 90; and health, 11, 20, 195; industrial, 24, 42, 72, 207n1; wastewater, 59. *See also* sewage; toxic water
pompanisasi (community-run water pump group), 2, 88, 98, 134, 136
postcolonial: engagement, 170, 172; infrastructure, 14, 18; reality, 6, 17, 36, 88, 90, 166
poverty, 13, 23, 36, 64, 100, 153. *See also* hardship; Kampung Improvement Program (KIP); PNPM Mandiri (National Program for Community Empowerment)
Povinelli, Elizabeth, 12, 89, 92, 93, 101
power hierarchies, 8, 126, 173, 180, 182
pride, 9, 60, 95, 103, 122, 152; airport, international, 6; "rent to nature," 121
Prihadi, Hendrar (Semarang mayor), 151–52, 179. *See also* mayoral visits
private donations, 14, 23, 24
problem zones, 28, 43, 62, 194
protest, 75, 103, 129, 141, 169, 170; Suharto (Indonesian president 1967–98), 128
provincial level, 29, 66, 85, 126, 171
PSDA (Semarang's water agency), 143, 144, 155, 168, 176, 183
public meetings, 123, 169

pumping house (polder), 157, 158, 179, 182, 183, 191; construction, 171, 177, 178, 181; eviction, 170, 181
pumps, polder, 176, 178, 180, 181; East Flood Canal, 186, 191; pumping house (polder), 157, 158, 179, 183, 191

"quasi-events" (discussion), 89, 92, 93, 101, 119, 194

racial differentiation, 28, 54
Rademacher, Anne, 135
railroad system, 16, 20, 27, 33, 41
rainfall, 38, 56, 92, 102, 157, 171
raising (*peninggian*), 3, 8, 106, 114, 145, 194; floors, 95, 98, 118; houses, 2, 34, 107–8, 111, 115; income, sources of, 187; infrastructure, 30, 126, 139, 153; and pride, 9; roads, 138, 149; roofs, 40, 116, 192; streets, 4, 11, 119, 121, 124, 132
"red" city (*kota merah*), 17
reforms, political, 22, 30, 31, 126, 150, 173
religion, 24, 38, 76, 88, 99
relocation, 5, 46, 54, 74, 102, 186
Rendy, 65–66, 67, 83
"rent to nature," 8, 122, 138–39, 149, 150, pride, 121
repair, constant, 7, 98, 135, 157, 194; chronic present, 5, 9, 87; grants, 14; house adaptations, 124; "meantime," 5, 120; *pompanisasi* (community-run water pump group), 2. *See also* house adaptations; raising (*peninggian*)

reproduction (of), 4, 9, 14, 99, 150, 153; capitalist reproduction, 10, 163; society, 13, 136, 162, 187; structures, 122, 126
research methods, 25, 64, 208–9n4; archival research methods, 19, 36, 43; interview methods, 104, 144, 212n10; participant observation, 61, 90, 110, 123, 177; photo-ethnographic methods, 90, 109. *See also* ethnographic methods
residential council (BKM). *See* BKM (Badan Keswadayaan Masyarakat)
resistance, 11, 16, 75, 170
responsibility, resident, 13, 96, 100, 149–50, 162; chronic present, 5, 11; criminality, 67, 68; ecological deterioration, 150; house adaptations, 124; labor, "community," 22, 99; RT/RW system, 23; state–society relations, 160; water tax, 161
retention basin (polder), 52, 156, 180, 183, 193; eviction, 139, 157, 181, 184
retribusi (informal land tax), 73
Rianto, Pak, 51–53, 184
river normalization. *See* normalization (*normalisasi*), concept of
river walls, 76, 93, 106, 133, 142
roads, inspection, 55, 68, 70, 73
rob, effects of, 4, 39, 56, 97, 131, 135; agency, 195; apocalyptic potential, 13; ecological deterioration, 150; endangered, 196; eviction, 7; and inequality, 5, 132; *pompanisasi* (community-run water pump group), 136. *See also* house adaptations

Roitman, Janet, 13, 162, 164, 169
Rozi, Pak, 52
RT/RW system, 21–22, 23, 24
Rückert, J. J., 48, 50
runoff, 2, 98, 134, 180, 191

sea level rise, 7, 29, 131, 142, 157, 161; apocalyptic potential, 13; delta, 158; littoral, 78; relocation, 5; subsidence, land, 167, 187
seasonal flooding, 26, 35, 56, 99, 157, 195
seeping riverbanks, 2, 4, 5, 120, 133
segregation, 13, 53, 153
self-governance, 5, 29, 90, 99, 120, 132
Semarang, municipality of. *See* municipality of Semarang
"Semarang Surga Yang Hilang," 86
sewage, 2, 48, 90, 98, 111
sewers, 46, 61, 111, 193
Siegel, James, 61, 64–65, 210n7
SIMA (water authority), 153, 160, 173, 183, 185; democratic reform, 154; eviction, 181; governance, participatory, 188; Indah, 55, 56, 172, 192; meetings, 172, 175, 177, 183–84, 186; office, 159, 186, 192; PSDA (Semarang's water agency), 168, 176; pumping house (polder), 171, 177, 178, 181, 182, 183
skepticism, 14, 41, 192
slums, 7, 18, 19, 24, 48, 53
social change, 4, 119, 131, 155
social control, 29, 69, 126
sosialisasi (community events), 27, 128, 140, 178
spatial hierarchy, 3, 6, 28, 54, 153
stacking, 5, 30, 91, 93, 126, 153

state development grants, 29, 126, 138
state intervention, 9, 157, 169, 170, 180, 206–7n6; colonial government, 47, 49; criminality, 28; inspection, 63; *kampungs*, formation of, 20, 46; normalization (*normalisasi*), concept of, 28, 69, 70; roads, inspection, 47, 73; SIMA (water authority), 153; skepticism, 14. See also *gotong-royong* (mutual assistance)
state–society relations, 70, 90, 124, 160, 177
stigmatization, 47, 63, 91, 195, 205n2
Stoler, Ann Laura, 47, 49, 51
strokes, 50, 51, 100, 110
subdistrict government (*kelurahan*), 22, 73, 91, 137, 139, 140; administration, 24; bridges, 71; grants, 18, 29, 126; New Order, 52, 127; office, 57, 122–23; *sosialisasi* (community events), 128; taxation, 180. See also *faskel* (PNPM facilitator); LPMK (Neighborhood Community Empowerment Board); *lurah* (subdistrict head)
subsidence, land, 6, 30, 78, 125, 161, 163; climate change, 29, 157, 196; debt, 105; groundwater extraction, 6, 72, 158, 196; house adaptations, 113; long-term thinking, 112; sea level rise, 167, 187
subsystem survey, 42, 173, 174
Suharto (Indonesian president 1967–1998), 70, 72, 130, 173, 209n4, 210n7; identity, Javanese, 17; New Order, 28, 52, 64; *pembangunan* (development), 8; protest, 75; social control, 69; *sosialisasi* (community events), 128
Sukawi Sutarip (Semarang mayor 2000–2010), 173
Sullivan, John, 21, 22, 177, 205–6n4
Sumarmo, 171, 175, 177, 187, 188, 192; eviction, 184, 186; power hierarchies, 182; retention basin (polder), 183, 193; *sosialisasi* (community events), 178; subsystem survey, 173, 174
suppression, 65, 127, 188
surveys, 147, 167, 170, 173, 174

talut. See embankments (*talut*)
Taussig, Michael, 18, 33, 37
taxation, 8, 43, 52, 176, 180, 196; *iuran* (community tax), 98, 134, 177; *retribusi* (informal land tax), 73; water tax, 81, 159, 161
technofix mentality, 4, 30
Tillema, H. F., 48, 49, 50
top-down development, 60, 72, 155, 184
toxic water, 2, 3, 18, 42, 86, 111
transparency, 28, 63, 124, 139, 143
"treading water," 4, 9

unfinished improvement works, 76, 82, 143, 186, 187
urbanization, 16, 37, 43, 82, 154, 195; dispossession, 193; ecological deterioration, 41; land, access to, 28; mangroves, 55
urban redevelopment plans, 5, 16
urban research methods, 25, 174, 208–9n4

Versnel, Hans, 17, 50
violence, 51, 62, 67, 89, 209n4, 210n7; colonial government, 50; domestic violence, 137; killings, mass, 16, 64, 65; marginalization, 61, 65; New Order, 64, 65, 136; "normalization of violence," 110, 163–64, 188
voluntary work (*sukarela*), 11, 21, 93, 96, 110. *See also* labor, "community"

wage labor, 20, 37
Wahyu, 27, 132, 137, 139, 145–49; Aliansi (Civil Alliance Against Flooding), 136, 143, 145; Komayu (resident organization), 140, 141, 142, 150; *lurah* (subdistrict head), 123–24; PNPM Mandiri (National Program for Community Empowerment), 146; "rent to nature," 138
Wahyudi, Imam (water expert), 172
warung (canteen), 90, 91, 100, 134–35; Arief, 96, 102–3, 110, 135; Ariel, 1, 96, 102
wastewater, 3, 42, 59, 72, 134, 191; *pompanisasi* (community-run water pump group), 2
water agency, Semarang's (PSDA), 143, 144, 155, 168, 176, 183
waterborne diseases, 2, 3, 46, 50
water governance, 61, 125, 166, 172, 179, 185; "multi-stakeholder" approach, 152, 153; seeping riverbanks, 133; and surveys, 174; technofix mentality, 4. *See also* SIMA (water authority)
water tax, 81, 159, 161
West Flood Canal, 15
World Bank, 29, 123, 129, 150, 176

Lukas Ley is an associate of the Max Planck Institute for Social Anthropology and lectures at the Center for Asian and Transcultural Studies, Heidelberg University.

Lightning Source UK Ltd.
Milton Keynes UK
UKHW021824310722
406641UK00003B/320